90 Advances in Polymer Science

Conducting Polymers/ Molecular Recognition

With contributions by
N. C. Billingham, P. D. Calvert, Y. Kurimura,
A. A. Litmanovich, I. M. Papisov

With 42 Figures and 9 Tables

Springer-Verlag Berlin Heidelberg GmbH

ISBN 978-3-662-15079-5   ISBN 978-3-540-46157-9 (eBook)
DOI 10.1007/978-3-540-46157-9

Library of Congress Catalog Card Number 61-642

© Springer-Verlag Berlin Heidelberg 1989

Originally published by Springer-Verlag Berlin Heidelberg New York in 1989
Softcover reprint of the hardcover 1st edition 1989

2152/3020-543210 — Printed on acid-free paper

Editors

Table of Contents

Electrically Conducting Polymers —
A Polymer Science Viewpoint

N. C. Billingham
School of Chemistry and Molecular Sciences, University of Sussex, Brighton BN1 9QJ, England

P. D. Calvert
Arizona Materials Laboratories, University of Arizona, Tucson, AZ85712, USA

During the development of conducting polymers over the last 10 years attention has tended to focus on the remarkable electronic properties of these materials. The fact that they are mainly very rigid-chain polymers also makes them remarkable as polymers. We review the synthesis, characterization and properties of conducting polymers and emphasis e comparisons with the flexible-chain materials such as the polyolefins.

The infusibility and insolubility of most conducting polymers prevents their being processed. This puts great emphasis on precursor routes where a soluble polymer can be processed to a film or fibre and then converted thermally to a conducting polymer. The intractability also limits characterization of molecular weight and chain microstructure. The morphology of conducting polymers ranges from highly crystalline to totally amorphous, depending on the material and the synthetic route. Until recently it has seemed that molecular weight and morphology had little influence on the conducting properties. New results on polyacetylene suggest that large increases in conductivity may be expected as highly oriented and perfect structures are achieved. These factors will also be of increasing importance as attempts to use these materials increase the interest in mechanical properties.

Applications of these materials will be dependent on their long term stability. The highly conjugated structures make conducting polymers very vulnerable to degradation. Mechanisms and kinetics of thermal and oxidative degradation are discussed. The current status of a number of applications

Advances in Polymer Science 90
© Springer-Verlag Berlin Heidelberg 1989

1 General Introduction

For most of the history of polymer technology one of the most valued properties of synthetic polymers has been their ability to act as excellent electrical insulators, both at high voltages and at high frequencies. In spite of this there has been interest for many years in the possibility of producing electrically conducting polymers. The obvious attraction is to combine in one material the electrical properties and high added-value applications of a semiconductor or a metal with the advantages of a polymer. The attractions of polymers to the electrical/electronics industries include the ease and low cost of their preparation and fabrication, as compared to semi-conductors and metals, especially in films and fibres, and their mechanical properties, particularly flexibility and impact resistance. For the polymer industry the high added-value of typical semiconductor materials represents a major attraction.

In reviewing conducting polymers, we will assume that there is some agreement on what is meant by "conducting" and concentrate on the meaning of "polymer". According to the Concise Oxford Dictionary, a polymer is a compound formed by simple chemical addition from a number of identical molecules each of which consists of a number of identical units. Many chemists would dispute the term "simple" and the rest of the definition is rather obscure, though clearly in the right spirit. A better definition of a polymer would be a material formed by the predominantly linear addition or condensation of similar units. In adopting this definition we would exclude the inorganic oxide networks and possibly also heavily cross-linked systems such as epochy resins.

Although the search for "organic metals" has followed many routes, we shall be concerned only with those polymers in which electrical conductivity arises mainly from the presence of sequences of conjugated carbon-carbon double bonds. It has long been known [1] that conjugated organic molecules can exhibit semiconductor properties. Early development was inhibited by the fact that the chain stiffness implicit in a conjugated structure also produces extreme intractability so that most of the early examples of conducting polymers were infusible, insoluble black powders of little value for either research or development. Two factors combined to change this situation dramatically. In the early 1970s Shirakawa and Ikeda [2,3] demonstrated the possibility of preparing strong, self-supporting films of polyacetylene by direct polymerization of acetylene. The polymer thus produced is a poor semiconductor and attracted little attention until 1977, when MacDiarmid et al. [4] rediscovered the fact that treatment of polyacetylene with Lewis acids or bases can increase its con-ductivity by up to 13 orders of magnitude. This process involves removal or addition of electrons to or from the polymer chain and is termed "doping" by analogy with silicon technology, although the analogy is poor.

Since the publication of the original observation in 1977 there has been an explosive growth of research into the whole range of conjugated polymer structures which has led to the development of a new family of polymers which, with appropriate chemical modification, can display conductivities from poor semiconductor to comparable with copper. These polymers have obviously been the result of major contributions from synthetic chemists and have been of great interest to solid-state physicists. Somewhat surprisingly there has been less interest on the part of polymer scientists. This has led to problems in many areas, partly because of failure to recognise the limitations of chemical synthetic methods when applied to polymers and partly because of a tendency to apply models based on perfectly crystalline, infinitely conjugated chains to amorphous and highly defective samples. In this review we aim to look at the synthesis, structure and conduction properties of conjugated organic polymers from the point of view of the polymer scientist.

Wegner [5] has argued that conducting polymers should be viewed as a simple extension of conducting molecular crystals in terms of their structure and conductivity. We would support this, in so far as the conducting polymers do not fall within the mainstream of polymer science. However, this is changing as synthetic methods for conducting polymers improve and as polymer science becomes more concerned with rigid, rather than flexible, chains.

In this context, we will be addressing a number of questions, which include:

a) Can conducting polymers be made with high molecular weight as defect-free chains, without cross-linking?
b) How do the structures and morphology compare with the amorphous, random coil or semicrystalline states of linear polymers?
c) How do the mechanical and other properties compare with "true" polymers?
d) How does the conductivity depend on the chain microstructure and conformation?

2 Polymer Structures and Synthetic Methods

The synthesis of a polymer requires the controlled coupling of a large number of monomer molecules to form the polymer chain. An obvious requirement for the monomer is that it shall be functionally capable of coupling to two other units and polymer chemists have long recognised three basic ways in which the required difunctionality can be achieved.

Opening of a carbon-carbon double bond is the most widely practised technology, typified by the large — scale production of polyethylene, polypropylene and the many other polymers with carbon backbones. In this case the monomer units are coupled in a chain — reaction polymerization, involving repeated addition of the olefin monomer to an appropriate active centre, which may be an ion, a free-radical or a coordination complex. Reactions of this kind can lead to polymers of very high chain length although the chain structures are often difficult to control, so that branching and stereochemical defects are common as are low concentrations of other defects caused by "wrong" addition of a monomer unit.

The second common method of polymer synthesis involves the stepwise coupling of small molecules which are difunctional by virtue of reactive functional groups. A typical example of step-reaction polymerization would be the synthesis of polyamides by reaction of a diamine with a diacid. In these systems the chain is built up slowly by reaction of any pair of functional groups in the system and it is common for the coupling to involve elimination of a small molecule. Conventionally these polymerizations allow more control over the chain structure but difficulties in reaching very high conversions and problems of reagent purity usually lead to much shorter chain lengths than can be obtained in addition polymerization. Polymers produced by this mechanism usually have functional groups forming part of the backbone.

The third major method for achieving difunctionality involves the ring-opening polymerization of a cyclic monomer, typified for example by the synthesis of polyamides from cyclic lactams. Reactions of this type proceed by chain-reaction mechanisms but yield polymers more typical of step-reactions, in that they contain functional groups within the chain.

Although direct synthesis from the monomers is the commonest way of producing polymers, there is also the possibility of using chemical modification of existing polymers to produce otherwise intractable or inaccessible structures. A major example of this type would be the production of poly(vinyl alcohol) by hydrolysis of poly(vinyl acetate). This method has been much favoured by chemists trying to synthesise highly intractable polymers, the approach being to produce a processable "precursor" polymer, which can be converted into the final structure after processing into the appropriate form. A large-scale example of this technology is the production of polyimide films via an intermediate polyamic acid. Where volatile products are released in the transformation process this method is restricted to thin films and fibres.

In the search for potentially conducting conjugated polymers all of these methods have been explored. In the following sections we survey the main types of conjugated structure and briefly outline the strategies which have been adopted to prepare them.

2.1 Conjugated Polymers by Addition Polymerization

2.1.1 Polyacetylene

Polyacetylene is the archetype of all conducting polymers and has attracted more attention than any of the other structures partly because it is the simplest conjugated structure and partly because the degenerate ground state of the all-trans-isomer allows the existence of mobile neutral spin defects ("solitons"), whose behaviour is of great interest to solid-state physicists. Polyacetylene has been extensively reviwed by Chien [6]. It was first synthesised as an intractable black powder by Natta et al. [7], in 1958, using catalytic amounts of the Ziegler-Natta catalyst $Ti(OBu)_4/AlEt_3$ in a conventional stirred reactor. The breakthrough for conducting polymers research came when Shirakawa et al. [2, 3, 4, 8] found that exposure of acetylene gas to high (10 mM) concentrations of the catalyst in a non-stirred reactor leads to the production of thin, strong films of polyacetylene on the walls of the vessel and the surface of the liquid. This polymerization has been extensively studied and is reviewed in great detail by Chien [6]. As might be expected it proceeds via the repeated cis-insertion of acetylene into a Ti-C bond and the initial product of reaction at low (−78 °C) temperature is a crystalline, predominantly cis-polymer I (Fig. 1).

^{13}C NMR shows that the C=C double bonds are located between the carbon atoms of the original acetylene molecule, showing that the polymerization is indeed a simple cis-insertion [9]. As prepared by this method polyacetylene is a free-standing film. On closer examination its density is found to be around 0.4 g cm^{-3}, only about 30 % of the value (1.16 g cm^{-3}) predicted from x-ray analysis, and electron microscopy

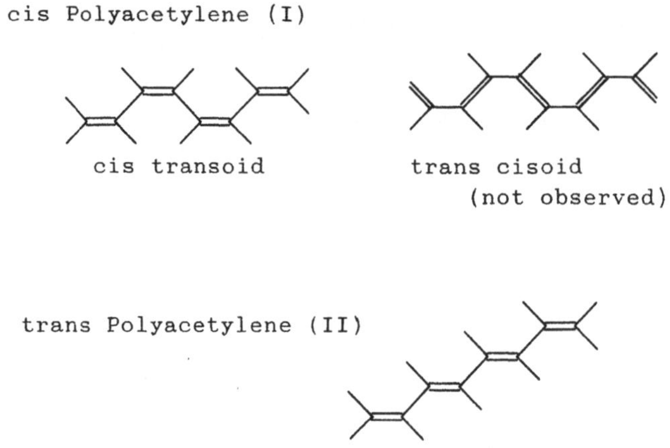

cis Polyacetylene (I)

cis transoid   trans cisoid
                               (not observed)

trans Polyacetylene (II)

Fig. 1. Possible geometric isomers of a polyacetylene chain

reveals complex open, fibrillar morphologies. The conductivity depends a great deal on the preparation conditions but the polymer is typically a poor insulator, with a conductivity (σ) of around 10^{-10} S cm^{-1}.

Although the *cis*-isomer is the metastable product of polymerization, increased reaction temperatures or post-polymerization heating of the *cis*-polymer lead to production of the thermodynamically more stable all-trans-form (II, Fig. 1). The isomerization is an interesting example of a solid phase reaction of a polymer. It requires rotation about C—C bonds in the polymer backbone, a very difficult process in a crystal lattice. Because of this, isomerization is associated with the formation of residual defects in the polymer, mainly in the form of unpaired electrons. These defects lead to a significant increase in conductivity in the trans-form, which is typically a poor semiconductor ($\sigma = 10^{-6}$ S cm^{-1}), although the conductivity is very dependent on preparation conditions. During isomerization the C=C double bonds become randomized among the carbon atoms [10].

Despite the popularity of the Shirakawa catalyst for acetylene polymerization, other initiating systems are capable of initiating the reaction and many are reviewed by Chien [6] and by Saxman et al. [11]. For example Cao et al. [12] have described the use of catalysts based on rare earth complexes, particularly neodymium phosphites. Not all catalysts produce the open fibrillar structure typical of Shirakawa polyacetylene. Catellani et al. [13], synthesised polyacetylene films using catalysts based on Ti(OiBu)$_4$ and aluminoxane co-catalysts. Their polymers have a fibrillar morphology but are much more closely packed than the normal Shirakawa material, giving densities in the range 0.6 to 0.8 g cm^{-3} and a specific surface area of the order of 50 m^2 g^{-1}. It was claimed that the polymers are stable enough to handle in air but no details of conductivity or doping were given. It is also claimed that the polymer can be extended to a draw ratio of 3.5. Recently, Kminek and Trekoval [14] have described the polymerization initiated by catalysts in which the titanium is reduced by Grignard reagents rather than aluminium alkyls. They claim that the films produced are much more compact, having densities in the range 0.9 to 1.05 g cm^{-3} and surface areas around 0.3 m^2 g^{-1}. There is evidence for some saturated groups in the polymer but it has good mechanical properties and can be doped with iodine to conductivities comparable with Shirakawa polyacetylene.

Theophilou et al. [15, 16] recently described the use of a highly viscous, complexing solvent (silicone oil) to control the morphology of polyacetylenes prepared on a series of catalysts derived from titanium alkoxides or titanium tetrabenzyl. They showed that the combination of the solvent and a post-polymerization thermal treatment in the oil gave fibrillar morphologies which were more compact than those from the typical Shirakawa system, with densities up to 1.0 g cm^{-3}. Using this approach, Naarman and Theophilou [17] report polymerization of acetylene to give a film which was claimed to have a much lower level of sp^3 defects and could be oriented to a draw ratio of 6.5. Basescu et al. [18] showed that this material can be doped with iodine to a conductivity of 10^4 S cm^{-1}, the highest value yet reported, and comparable to the highest metallic range.

A different method of polymerizing acetylene is the use of metathesis catalysts, in which the chain propagation involves insertion into a transition-metal carbene complex [19]. Amass et al. [20] reported that typical methathesis catalysts, such as WCl$_6$/AlEtCl$_2$, metathesise acetylene to the polymer in only rather poor yields. Theophilou et al. [21] reported that the catalyst system derived from WCl$_6$ and n-BuLi polymerizes acetylene to yield a high (70%) $trans$-polymer at room temperature; the polymer has a globular morphology and is doped by the WCl$_6$.

One of the most interesting alternatives to the Shirakawa catalyst has been the systems disclosed by Luttinger [22, 23] and later elaborated by Lieser et al. [24]. The tris(2-cyanoethyl)phosphine complex of nickel chloride reacts with sodium borohydride to produce a catalyst system capable of polymerizing acetylene in solutions in either alcohol or, quite remarkably, water. A more efficient catalyst is obtained by replacing the nickel complex with cobalt nitrate. Interest in Luttinger polyacetylene seems to have waned in the last few years.

Some attempts have been made to synthesise polyacetylene by other routes with less success. Korshak et al. [25] tried to produce polyacetylene by the metathesis of cyclooctatetraene; although they obtained deeply coloured, conjugated products these were mainly of low molecular weight and there is evidence of side reactions. In contrast, Klavetter and Grubbs [26] used a preformed tungsten carbene complex to polymerize cyclooctatetraene and obtained high yields of a high $trans$-polymer. The undoped conductivity of their product was 10^{-11} S cm^{-1}, almost six orders of magnitude lower than that of the Shirakawa polymer, but the material could be doped to similar conductivities on exposure to iodine. Heaviside et al. [27] showed that acetylene polymerizes on contact with the surface of gamma alumina and Murakami et al. [28] showed that this process can be accelerated by milling the alumina in an acetylene atmosphere to induce mechanochemical initiation.

2.1.2 Substituted Polyacetylenes

2.1.2.1. Polymers by Direct Polymerization

The polymerization of mono- and di-substituted acetylenes has been very extensively investigated, using all possible methods for catalysis of addition polymerization. Alkyl, aryl, trifluoromethyl and halogen substituents have all been reported and the polymerizations have recently been reviewed by Feast[29] and by Gibson and Pochan[30]. Research in this area has been disappointing from the point of view of conducting polymers. Although many of the polymers are well enough characterized for us to be sure that they are true polymers with long conjugated sequences, few of them exhibit conductivity. Many are white or pale yellow solids, soluble in organic solvents and showing no response to doping. This is due partly to the fact that steric hindrance along the polymer chains limits the extent of p-overlap, partly to the fact that the substituents also force the chains to lie further apart in the solid, so limiting π-overlap between chains and partly to the fact that electron-donating or-accepting substituents can increase the non-equivalence of the carbon atoms, forcing higher bond alternation and a higher band gap.

Polyphenylacetylene is an orange solid which is an insulator ($\sigma = 10^{-12}$ S cm^{-1})[31, 32]. Although it has been reported to dope with I_2 and with AsF_5 to conductivities around 10^{-5} S cm^{-1}, the nature and reversibility of the doping is rather uncertain[33] and it has been suggested that the conductivity is ionic[34]. Recently, Wentworth and Bergquist[35] following earlier work by the US Army, looked at the conductivities of a range of substituted phenylacetylene polymers, hoping to find specific chemical responses useful for chemically selective electrodes. They prepared their polymers by thermal polymerization of highly purified monomers and found that all of them were electrical insulators. They concluded that the conductivity observed earlier arises from catalyst residues.

Polymethylacetylene can be doped by I_2, but not by AsF_5. Its conductivity in the doped state[36] is around 10^{-3} S cm^{-1}. As might be expected, increasing content of mono- and di-substituted acetylenes in copolymers with acetylene gives increasing solubility at the expense of drastic reductions in the conductivity of both doped and undoped polymers.

Quite recently[37] Zeigler used the metathesis catalyst based on WCl_6 and n-BuLi to polymerize trimethylsilylacetylene to a soluble polymer with a styrene-equivalent M_n of around 10^5. The polymer was soluble, stable in air and dopable with I_2 to a conductivity of 10^{-4} S cm^{-1}. It was suggested that the reactivity of the $SiMe_3$ group would allow access to a whole range of new functionally substituted polymers but little detail has yet been published. It seems possible that the conductivity of the poly-

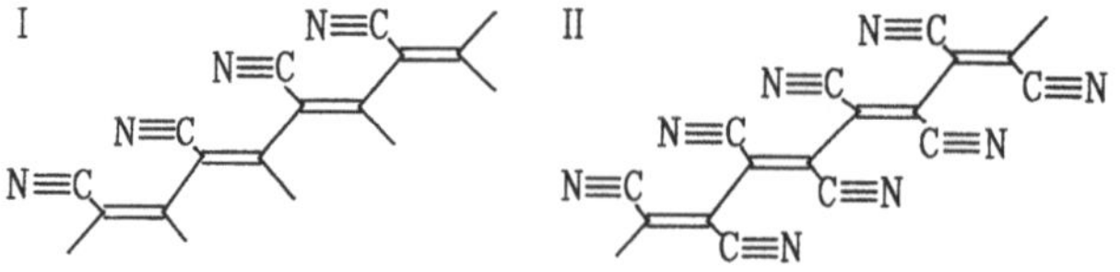

Fig. 2. Polycyanoacetylene

mer is ionic since all of the I_2 can be pumped out, returning the polymer to its undoped conductivity.

Some estimate of the relative importance of steric and electronic effects in substituted polyacetylene can be gained by studying the polymers of 1-cyanoacetylene and 1-2-dicyanoacetylene (Fig. 2). Structure I has the polar, electron-withdrawing cyano-group on alternate carbon atoms, which would be expected to lead to a very high degree of bond alternation and high band gap. In contrast, poly(dicyanoacetylene) (II) has the same symmetry as polyacetylene and would be expected to have a similar, weak, bond alternation. The synthesis of both of these polymers has been reported by Chien and Carlini [38] who describe extensive studies of paramagnetism and conductivity. The surprising fact is that both polymers behave much like undoped *cis*-polyacetylene, showing immobile defects and low conductivities. Both polymers will dope only to very low conductivity and the dopant can be removed completely by pumping. These observations suggest that steric effects, causing twisting of the chain and poor conjugation, play a much more important role in determining conductivity than do electronic effects. Leclerc and Prud'homme [39] used extended Huckel calculations to show that the band gap in substituted polyacetylenes is expected to be very sensitive to twisting of adjacent repeat units away from the planar, conjugated structure, and Bredas et al. [40] drew similar conclusions for aromatic polymers.

MacDiarmid and Heeger [41] reported that polyacetylene itself can be brominated then dehydrobrominated to yield a polymer with about 20% of the hydrogen atoms replaced by Br. The product is a semiconductor ($\sigma = 10^{-5}$ S cm^{-1}) which can be doped to high conductivity. Gibson [42] also reports that treatment of polyacetylene with SO_3 gives a (presumably sulphonated) product with a conductivity of 10^{-1} S cm^{-1}. It is not clear in either of these studies whether the polyacetylene reacts uniformly or whether reaction is confined to amorphous regions. In view of the high reactivity of many of the dopants used in polyacetylene, particularly in the earlier stages of its development, it is probable that many of the polymers studied were in fact partially substituted by reaction with the counter-ions.

2.1.2.2 Cyclopolymerization as a Route to Polyacetylenes

Although the polymerization of diene monomers is most familiar for 1,3-dienes, as in the production of rubbers, the polymerization of 1,6-dienes to yield polymers containing six-membered rings ('cyclopolymerization') has been well established for many years [43]. Gibson et al. [44] have used cyclopolymerization of 1,6-diynes to prepare polymers which are effectively substituted polyacetylenes, the archetype being the polymerization of 1,6-heptadiyne:

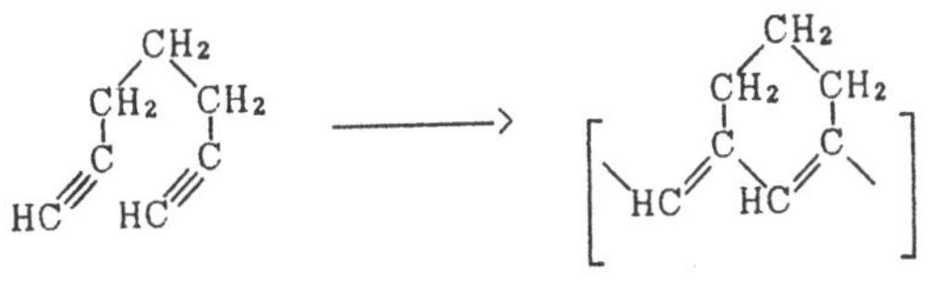

This reaction was described by Gibson et al. [44] and the properties of the polymer have been reviewed [45]. Ziegler polymerization gives the cyclopolymer as a soluble material which can be converted into conducting films that can be doped with I_2 to high (1 S cm^{-1}) conductivity. Unfortunately the polymer is not stable, undergoing a 1,3-hydrogen shift on heating or doping to give the more stable but non-conjugated isomer:

This isomerization is accompanied by a drop in conductivity. The polymer is also unstable in air, oxidising more rapidly than does Shirakawa polyacetylene. Both of these problems result from the allylic hydrogen atoms in the 3,5-positions of the monomer. In an effort to remove these hydrogens Gibson looked at at the polymer derived from propiolic anhydride:

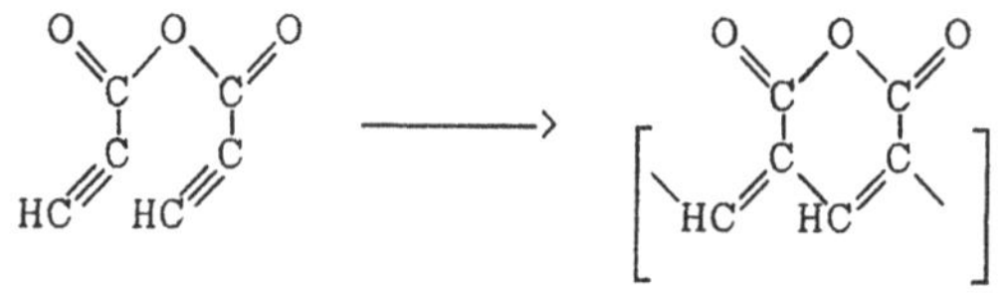

This polymer was readily synthesised as a soluble film-forming material, dopable with I_2 to a conductivity around 10^{-4} S cm^{-1} and of good oxidative stability. However, it was of low molecular weight and contained copolymerized fragments derived from reactions with the solvent. In trying to overcome these problems, Gibson and Weagley [46] have looked at a large number of other cyclo-monomers. Despite some elegant synthetic work, none of the desirable monomers was synthesisable in sufficient purity to be usable.

2.1.3 Acetylene Copolymers

In polymer science, copolymerization is commonly used to obtain properties in a polymer system which are intermediate between those of the two homopolymers. Since polyacetylene is a good conducting polymer and the substituted polymers are generally poor conductors, it might be expected that copolymers of acetylene with its substituted analogous would be intermediate in conduction properties and this is indeed generally true. Chien et al. [6, 36] used Shirakawa catalysts to make copolymers of acetylene with methylacetylene, the latter being chosen because the homopolymer is soluble and the methyl group was expected to produce minimal distortion of the electronic structure. Copolymers with up to 80 mol% of methylacetylene were not soluble in common solvents although they could be swollen. The copolymers could be doped with iodine or AsF$_5$ but the limiting conductivities were lower by up to 6 orders of magnitude in the copolymers. The copolymers were also found to be more sensitive than pure polyacetylene to both heat and oxygen and they had a larger band gap.

2.2 Polymers by Step-Reaction

Polymer synthesis by step-reaction involves the coupling of small molecules which are difunctional by virtue of having two reactive functional groups. It is common for the coupling to involve elimination of a small molecule, and typical examples might be the synthesis of a polyester from a hydroxy-acid or of a polyamide by reaction of a diamine with a diacid. Step-reaction polymerizations are often equilibria, whereas chain-reaction polymerizations are more usually irreversible.

In a chain-reaction polymerization, the polymer is formed by addition of monomer molecules to a relatively small number of propagating active centres. If the propagation is sufficiently rapid, high molecular weight chains can be produced even though only a small fraction of the monomer has been consumed. In contrast, in a step-reaction the chain is built up by reaction of any pair of functional groups and the chain length becomes long only if very high conversions can be achieved. This imposes very strict requirements on monomer purity since even small traces of monofunctional impurity will limit the maximum attainable molecular weight, whilst traces of poly-functional impurites can give branched or cross-linked structures. In almost all conducting polymers the polymer is insoluble in the reaction medium and precipitates. Where chain reaction is involved, further polymerization involves a monomer molecule diffusing to an active site in the solid polymer. In contrast, chain extension in a step-reaction polymer under precipitation conditions needs a much less favourable inter-action between two chain ends. For these reasons, step-reaction polymerization typically yields polymers with relatively short chain lengths, even under the most favourable conditions.

In principle, it should be possible to prepare polyacetylene and its derivatives by coupling reactions of appropriately 1,2-substituted olefins. In practice this route does not appear to have been explored and step-reactions have mostly been applied to prepare aromatic and heteroaromatic polymers. Some of the more important syntheses ae reviewed below; others have been reviewed by Feast [29].

2.2.1 Poly(p-Phenylene)

The synthesis of poly(p-phenylene) by step-reaction has a long history and has been reviewed [47, 48]. The Ullmann coupling of p-dihalobenzenes, catalysed by copper, and the Wurtz-Fittig coupling, in the presence of Na or K, both give oligo-phenyl-enes [49, 50]. Coupling of bis-diazonium halides has also been reported [51]; the product was a mixture of soluble oligomers which contained enough nitrogen to suggest the presence of diazo-bonds between at least some of the aromatic nuclei.

A much more successful approach to synthesis of poly(p-phenylene) from dihalo-benzenes is via the decomposition of the corresponding Grignard reagent:

The coupling can be catalysed by the presence of Fe(III) or Co(II) but is most efficiently induced by the dipyridyl complex of Ni(II) chloride, as described by Yamamoto et al. [52]. It is also efficiently promoted by 1,4-dichoro-2-butene, as described by Taylor et al. [53]. The structures of these "Yamamoto" polyphenylenes are discussed in Sect. 3; at this point it is sufficient to note that they are yellow/brown infusible powders, apparently having a high degree of p-coupling but relatively short (10—12 rings) chain lengths.

The alternative route to synthesis of poly(p-phenylene), which has been widely used, is the Scholl reaction [54], which involves direct oxidative elimination of two aryl hydrogen atoms with concomitant formation of a new carbon-carbon bond. This reaction occurs under Friedel-Crafts conditions and requires the presence of an appropriate oxidant to remove the hydrogen liberated in the coupling process. This route is typified by the "Kovacik" polymerization of benzene [55] induced by aluminium chloride in the presence of stoichiometric amounts of Cu(II) chloride:

$$ n \; \langle \bigcirc \rangle \; + \; 2n \; CuCl_2 \; \xrightarrow[H_2O]{AlCl_3} \; (-\langle \bigcirc \rangle-)_n \; + \; 2n \; CuCl \; + \; 2n \; HCl $$

This reaction is rapid and typically yields poly(p-phenylene) in the form of a brown powder, although Tieke et al. [56] obtained thin films with a granular morphology by polymerization in a strong shear field. Characterization suggests short chains with more irregularity than those formed by the Taylor or Yamamoto methods. Many other oxidants can be used in place of the $CuCl_2$, though less effectively. The functions of catalyst and oxidant can be combined in copper(II)tetrachloroaluminate [57] and in some cases a single molecule can double as catalyst and oxidant. Thus $FeCl_3$ and $MoCl_5$ can both induce polymerization of benzene [48, 58, 59, 60], though the products tend to be dark coloured, indicative of some doping. Shacklette et al. [61] showed that exposure of benzene and its oligomers to AsF_5 gives highly conductive polymers by simultaneous polymerization and doping and this approach has been extended to other aromatic and heteroaromatic polymers, including anthracene [62, 63], Aldissi and Liepins [64] reported that exposure of a solution of benzene in AsF_3 to AsF_5 gives a homogeneous solution containing poly(p-phenylene). None of these highly oxidized polymers is particularly well characterized and it is probable that they are highly irregular structures.

The mechanism of the polymerization of benzene on the $AlCl_3/CuCl_2$ catalyst has been the subject of discussion [65, 66]. Early papers suggested that it involves protonation of the monomer by the strongly acidic species formed by interaction of $AlCl_3$ with traces of water, followed by a cationic propagation reaction (Fig. 3a). It is now generally accepted that the first stage of the reaction is the electron-transfer oxidation of benzene to produce the radical cation (Fig. 3b). According to Kovacik et al. [48, 67] this species serves as the site for association of a series of benzene molecules to form a loosely complexed group with the radical cation character delocalized over all members. Oxidation of this complex by the $CuCl_2$ then gives the final polymer.

According to this view, the chain length of the polymer is limited by the delocalization of the reactive intermediates. The evidence for this rather unusual mechanism has been reviewed [48].

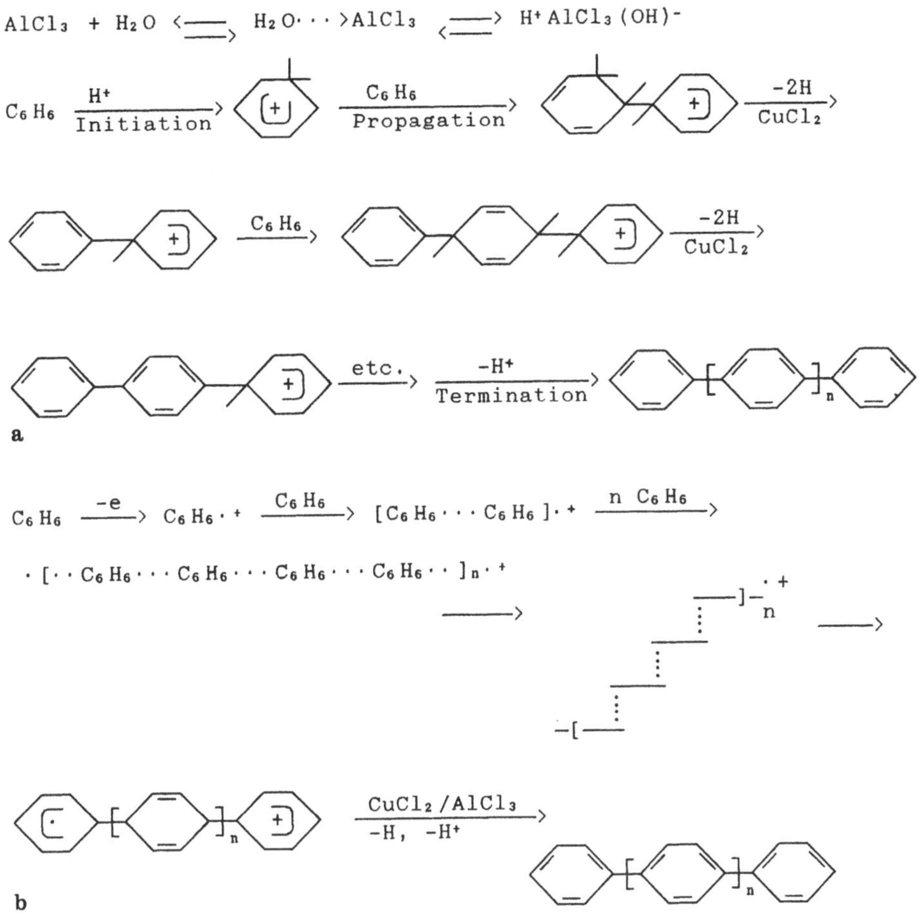

$AlCl_3 + H_2O \longleftrightarrow H_2O \cdots > AlCl_3 \longleftrightarrow H^+ AlCl_3(OH)^-$

$C_6H_6 \xrightarrow[\text{Initiation}]{H^+} (+) \xrightarrow[\text{Propagation}]{C_6H_6} \xrightarrow{-2H}{CuCl_2}$

$\xrightarrow{C_6H_6} \xrightarrow{-2H}{CuCl_2}$

$\xrightarrow{etc.} \xrightarrow[\text{Termination}]{-H^+}$

a

$C_6H_6 \xrightarrow{-e} C_6H_6 \cdot^+ \xrightarrow{C_6H_6} [C_6H_6 \cdots C_6H_6] \cdot^+ \xrightarrow{n\ C_6H_6}$

$\cdot [\cdots C_6H_6 \cdots C_6H_6 \cdots C_6H_6 \cdots C_6H_6 \cdots]_n \cdot^+$

$\longrightarrow$

$\xrightarrow[-H, \ -H^+]{CuCl_2/AlCl_3}$

b

Fig. 3a and b. Proposed mechanisms for the polymerization of benzene by the Kovacik route

2.2.2 Poly(p-Phenylene Sulfide)

One of the major problems with most conducting polymers is their intractability, resulting from the stiffness of their conjugated chains. One exception is poly(p-phenylene sulfide), which has a T_g of about 85 °C, a crystalline melting point at 280 °C and is soluble in organic solvents, such as diphenyl ether or chlorobenzene above 200 °C [68]. The polymer is produced commercially by the Phillips Petroleum Company, under the trademark 'Ryton'. Clarke et al. [69] reported that it can be doped with AsF_5 to give conductivities in the range 1 to 10 S cm^{-1}; the doping rate of thick samples is dramatically increased by simultaneous exposure to AsF_5 and AsF_3, the latter apparently acting as a plasticizer [70].

Synthetic routes to poly(p-phenylene sulfide) have been reviewed by Cleary [71]. The route used commercially is the polycondensation of 1,4-dichlorobenzene with sodium sulfide, typically carried out in a polar organic solvent, such as N-methyl pyrrolidone at temperatures in the range 200 to 300 °C:

$$n\ Cl - \langle\!\!\bigcirc\!\!\rangle - Cl\ +\ Na_2\,S\ \longrightarrow\ [-\langle\!\!\bigcirc\!\!\rangle-S-]_n\ +\ 2NaCl$$

This reaction is reported to be much more difficult to perform than it appears. According to Rajan et al.[72] the mechanism is not typical of polycondensations, since appreciable amounts of high polymer are present at low conversions and un-reacted monomer is found at high conversions. Numerous other synthetic routes have been described. According to Cleary[71] the only other process which has been unequivocally demonstrated to give the required polymer is the self condensation of alkali metal salts of 4-halothiophenols, as described by Lenz et al.[73].

$$n\ X - \langle\!\!\bigcirc\!\!\rangle - SM\ \xrightarrow{250^\circ C}\ [-\langle\!\!\bigcirc\!\!\rangle-S-]_n\ +\ MX$$

Although succesful, this reaction does not lead to polymers of sufficiently high molecular weight to be useful commercially.

The literature on conductivity in poly(p-phenylene sulfide) is confused. According to Shacklette et al.[74] heavy doping with AsF_5 causes reduction of the polymer with the formation of fused benzothiophene structures which are responsible for conjugation. This would more properly place poly(p-phenylene sulfide) in the category of precursor polymers, discussed later. On the other hand, Friend and Giles[75] proposed an intrinsic conduction mechanism, based on optical measurements and Tsukamoto et al.[76] have presented XPS and ^{13}C NMR measurements to support this view.

Poly(p-phenylene oxide) is also a processible polymer, but its treatment with AsF_5 leads only to rather poor conductivity[77]. The synthesis of poly(p-phenylene selenide) has been described by Jen et al.[78] and by Sandman et al.[79] but the polymer does not conduct even on prolonged exposure to AsF_5.

2.2.3 Heteroaromatic Polymers

Thiophene, pyrrole and their derivatives, in contrast to benzene, are easily oxidized electrochemically in common solvents and this has been a favourite route for their polymerization, because it allows in situ formation of thin films on electrode surfaces. Structure control in electrochemical polymerization is limited and the method is not well suited for preparing substantial amounts of polymer, so that there has been interest in chemical routes as an alternative. Most of the methods described above for synthesis of poly(p-phenylene) have been applied to synthesise polypyrrole and polythiophene, with varying success.

Chemical synthesis of polypyrrole has mostly been achieved by exposure of the monomer to strong oxidants, typically Fe(III)[80, 81], producing the polymer as a black powder. The polymer is formed and simultaneously oxidized to the doped state, with the anion included as the dopant counter-ion. By choosing the appropriate iron salt, it is possible to produce polymers containing a wide range of counter-ions[82, 83]. Polymers formed in this way have been studied as battery electrodes[84]. Simultaneous oxidation and doping of pyrrole has also been achieved using chlorine[85] or iodine or bromine[86] as the oxidant, although there is likely to be at least some addition of the halogen to the polymer chain. Organic electron acceptors, such as chloranil [87], have also been used. Chao and March [88] used a range of metal salts to produce polypyrrole in aqueous solution and found that all of the metal salts produced similar polymers, irrespective of considerable differences in their reduction potentials; in particular, they were unable to produce polymer without simultaneous oxidation to the doped state, even with the mildest oxidants. Bocchi and Gardini [89] have reported synthesis of polypyrrole films by reaction at a benzene-water interface.

The oxidative polymerization of pyrrole can also be induced in solvent-free systems and several papers have described synthesis via exposure of a matrix saturated with an Fe(III) salt to pyrrole vapour [90, 91]. Mohammadi et al. [92] described synthesis on planar substrates by sequential exposure to $FeCl_3$ and pyrrole and have also discussed a similar synthesis using H_2O_2 as the oxidant [93].

Although thiophene can be polymerized by exposure to oxidising agents, its higher oxidation potential is expected to give more reactive propagating centres and greater structural irregularity. For this reason, most chemical syntheses have used coupling reactions of Grignard compounds or similar systems. The typical reaction is the Ni-catalysed coupling of the Grignard reagent derived from 2,5-dibromothiophene [94, 95]. Kobayashi et al. [96] suggested that the chain length in polymers derived in this way is short, limited by impurities in the monomers and claimed to obtain better polymers by using milder reaction conditions and 2,5-diiodothiophene. Cunningham et al. [97] have presented a preliminary report on synthesis of thiophene and substituted thiophene copolymers via the Ni-catalysed Grignard coupling reaction.

A more recent report [98] proposed coupling of 2,5-dibromo-3-methylthiophene by reaction with butyl lithium, catalysed by $CuCl_2$. The authors later pointed out that this route is very critically dependent on the source of the copper catalyst [99] although Berlin et al. [100] have used the same chemistry to produce polymers containing thiophene units. Thiophenes alkylated in the 3-position can become soluble in organic solvents if the chain length of the alkyl group is long enough. These soluble polymers are discussed later; their synthesis has been achieved both by Grignard coupling and by oxidation with Fe(III) [101].

2.2.4 Wittig and Schiff's-Base Polymers

Although the methods reviewed above have been most widely used for synthesis of conducting polymers, there are many other organic reactions capable of producing conjugated structures. One example is the Wittig condensation of bis-triphenyl-phosphonium compounds with dialdehydes in the presence of a strong base, typified by the synthesis of poly(p-phenylene-vinylene) [102].

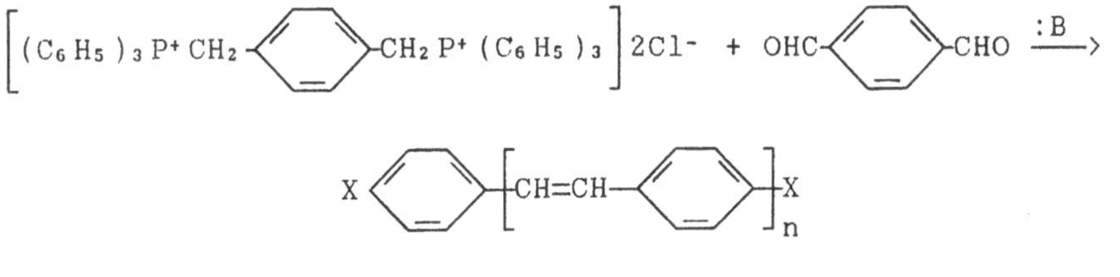

This route has been adapted by Kossmehl et al. [103, 104] to produce a wide range of conjugated polymers containing aromatic and heteroaromatic nuclei. Droske et al. [105] have prepared poly(arylenedifluorovinylenes) by reaction of tetrafluoro-ethylene with the dilithiated aromatic nucleus.

Two other related routes to conjugated polymers are the base-catalysed conden-sation of methyl-substituted aldehydes:

$$CH_3 RCHO \xrightarrow[-H_2 O]{KOC(CH_3)_3} [-RCH=CH-]_n$$

and the related polycondensation of a Schiff's base:

$$H_3 CRCH=NC_6 H_5 \xrightarrow[-C_6 H_5 NH_2]{KOC(CH_3)_3} [-RCH=CH-]_n$$

Both of these reactions have been adapted by Kossmehl et al. [103, 106] to produce polymers with aromatic and heteroaromatic units in the main chain. The advantage of these routes is the very wide range of polymer structures which is accessible; the main problem is that the polymers are formed as insoluble, infusible powders and are most probably oligomeric.

2.3 Electrochemical Polymerizations

Electrochemical initiation of polymerization reactions is not new and has been applied to the synthesis of a wide range of polymer types [107, 108]. In a conventional electro-initiated addition polymerization, the oxidation or reduction of an added soluble initiator takes place at the electrode surface to generate an active species, which may be an anion, a cation or a free-radical. Polymer formation takes place in solution and it is generally contrived that the polymer remains dissolved in the reaction medium, since the precipitation of a layer of insulating polymer on the electrode surface stops the reaction. The advantage of electrochemical methods of generating the initiating species is the degree of control which can be obtained over the rate of initiation and the concentration of active centres. Nevertheless, electrochemical methods are still laboratory curiosities for most insulating polymers.

Useful electrochemical synthesis of conducting polymers requires a combination of conductivity with insolubility in the electrolyte, which causes the polymer to preci-pitate, hopefully directly on the electrode surface, as a coherent film which does not passivate the electrode and may be removable from its surface for further study. Electrochemical synthesis is generally limited to rather small amounts of polymer, although a method of producing large amounts of film by continuous stripping of a rotating electrode has been described [109].

Kornicker [110] found that acetylene can be polymerized electrochemically using a Pt cathode, Ni anode and $NiBr_2$ as electrolyte in acetonitrile solution, the product being a black powder. Farafnov et al. [111] showed that phenylacetylene can be poly-merized electrochemically in dimethylsulfoxide solution, using $NiBr_2$ as the electrolyte. The reaction gives low yields of powdered polymer. Subramanian [112] showed that the polymerization is anionic and is accompanied by cyclization to give a proportion of cyclohexadiene rings in the chain. Chen and Shy [113] showed that this method can be adapted to produce thin films of polyacetylene on a Pt electrode. The films so produced were granular with a high surface area, typically $70 \ m^2 \ g^{-1}$, comparable to Shirakawa polyacetylene. The polymer is formed on the cathode and the authors speculate that the initial step is the formation of a complex between the acetylene and the nickel salt which is reduced at the electrode surface to give an anion which propagates the polymer chain. The possibility that the initiation is via an electrochemically generated nickel species was said to be excluded by the fact that no polymer was formed in the bulk of the solution until the electrode surface was completely coated with polyacetylene; whilst this is plausible, it is not totally convincing since slow generation of an active species followed by rapid polymerization would yield the same result.

Table 1. Structures and electrochemical properties of polypyrroles and polythiophenes

Monomer	Conductivity $S \ cm^{-1}$	Oxidation potential V vs SCE	Soluble?
$R=H$	40–100	−0.2	N
$R=CH_3$	10^{-4}	0.46	N
$R_1=H, R_2=C_2H_5$		0.15	N
$R_1=R_2=CH_3$	0.2–10	0.1	N
$R_1=R_2=C_2H_5$	10^{-4}	0.23	N
$R_1=H, R_2=(CH_2)_4SO_3Na$	0.01		N
$R_1=H, R_2=COCH_3$		0.9	N
$R_1=H, R_2=COC_{11}H_{23}$	360		Y
$R_1=H, R_2=COC_{17}H_{35}$	10		Y
$R_1=H, R_2=COC_6H_5$	10^{-3}	0.9	N
$R_1=R_2=H$	10	1.1	N
$R_1=H, R_2=CH_3$	100–750	0.9	N
$R_1=H, R_2=C_2H_5$	240		N
$R_1=H, R_2=C_4H_9$	110		Y
$R_1=H, R_2=i-C_4H_9$	2		N
$R_1=H, R_2C_6H_{13}$	95		Y
$R_1=H, R_2=C_8H_{17}$	78		Y
$R_1=H, R_2=C_{12}H_{25}$	67		Y
$R_1=H, R_2=C_{18}H_{37}$	17		Y
$R_1=H, R_2=C_{20}H_{41}$	11		Y
$R_1=H, R_2=C_6H_5$	100		N
$R_1=H, R_2=CH_2C_6H_5$	13		Y
$R_1=H, R_2=CH_2OCH_3$	0.31		Y
$R_1=H, R_2=CH_2O(CH_2)_2OCH_3$	51		Y
$R_1=H, R_2=CH_2O(CH_2CH_2O)_2CH_3$	1050		Y
$R_1=H, R_2=CH_2NHCO(CH_2)_{10}CH_3$	200		Y
$R_1=H, R_2=O(CH_2CH_2O)_2CH_3$	0.05		Y
$R_1=H, R_2=(CH_2)_2SO_3Na$	0.01		Y
$R_1=H, R_2=(CH_2)_4SO_3Na$	0.01		Y
$R_1=R_2=CH_3$	50	1.1	N
	10^{-4}		N

Virtually all of the real interest in electroinitiated synthesis of conducting polymers has focussed on the anodically active aromatic monomers, of which the most highly studied examples are pyrroles and thiophenes (Table 1).

These polymerizations depend upon the ability to oxidize the monomer to a radical cation, whose further reactions lead to polymer. Since the oxidation potentials of the polymers are lower than those of the corresponding monomer, the polymer is simultaneously oxidized into a conducting state so that it is non-passivating. Some of the more important electrochemically-synthesised structures are discussed in more detail below and Chandler and Pletcher [114] have reviewed the electrochemical synthesis of conducting polymers. Detailed discussion in terms of thermodynamic parameters is impossible because the polymerizations are irreversible, so that E_0 is undefined for the monomer-polymer equilibrium.

2.3.1 Polypyrrole

The chemical oxidation of pyrrole to produce an uncharacterized polymeric 'pyrrole black' has been known for many years but the polymer is non-conducting and contains oxygen. The electrochemical synthesis of pyrrole black as a brittle conducting film was reported by Dall'Olio et al. in 1968 [115]. With the renewal of interest in conjugated structures, the synthesis of polypyrrole has been reinvestigated, most notably in the laboratories of IBM [116]. The polymerization is reviewed by Diaz [117] and by Clarke et al. [118]. It takes place at the surface of an electrode (typically platinum or ITO glass) immersed in an aprotic solvent (typically acetonitrile) containing a dissolved salt, such as tetraethylammonium tetrafluoroborate. Polymerization apparently does not occur under rigorously anhydrous and oxygen-free conditions and some reducible species must be present to allow the cathode reaction to occur at sensible potentials; silver salts have been commonly used [119]. The mechanism of polymerization (Fig. 4) is generally believed to involve the oxidation of the monomer to yield a radical cation, so that the oxidation potential of the monomer determines whether initiation can take place.

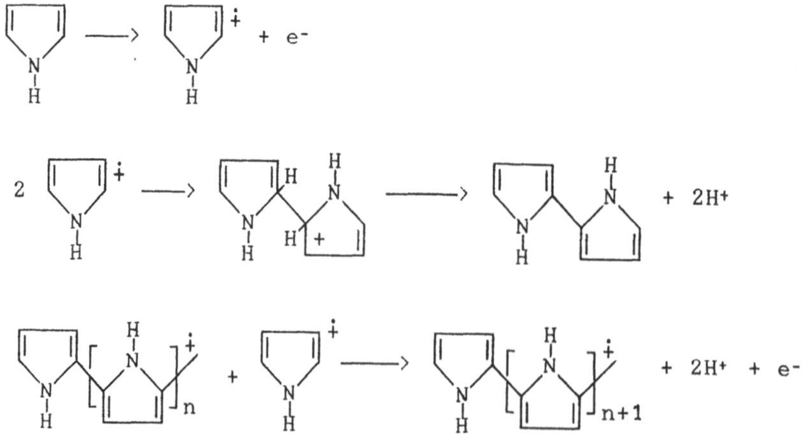

Fig. 4. Proposed mechanism of electrochemical polymerization of pyrrole

Provided that there are no nucleophiles in the system capable of reacting with the radical cations, they will couple to give a dimeric dication, which readily eliminates $2H^+$, rearomatising to give the pyrrole dimer. The oxidation potential of the dimer is slightly lower than that of the monomer so the process of oxidative coupling of the units continues as the polymer grows, becomes insoluble and precipitates on the electrode surface where growth apparently continues to give a high molecular weight polymer. In view of the very low diffusion coefficients of small molecules in the polymers, it is rather surprising that the molecular weight of the polymer is apparently not much limited by precipitation, as is so often the case with other polymerizations. The exact morphology of the polymer depends on the solvent composition and the current density, but it is typically a coherent film with a density of about 1.48 g cm^{-3}. Attempts to improve the structural regularity of the film by using α-bipyrrole as the monomer have been described [120].

Because the oxidation potential of the polymer is lower than that of the monomer, the polymer is electrochemically oxidized into a conducting state, kept electrically neutral by incorporation of the electrolyte anion as a counter-ion. This is an essential since precipitation of the unoxidized, insulating polymer would stop the reaction. Both coulometric measurements and elemental analysis show approximately one counter-ion per four repeat units. An important feature is the fact that the polymerization is not reversible whereas the oxidation of the polymer is. If the polymer film is driven cathodic then it is reduced towards the undoped state. At the same time neutrality is maintained by diffusion of the counter-ions out of the film and into the electrolyte. This process is reversible over many cycles provided that the film is not undoped to the point where it becomes too insulating. It is possible to use it to put new counter-ions into the film, allowing the introduction of ions which are too nucleophilic to be used in the synthesis. The conductivity of the film for a given degree of oxidation depends markedly on the counter-ion, varying by a factor of up to 10^5.

Pyrrole is sufficiently soluble in water to be polymerizable from aqueous solutions of suitable electrolytes [121], with the advantage that a much wider range of counter-ions is accessible. Warren and Anderson [122] and Qian et al. [123, 124] have reported a wide range of counter-ions. Of particular interest is the possibility of including very large counter-ions. Wernet et al. [125, 126] used aqueous solutions of detergents to prepare polypyrroles having counter-ions derived from N-alkyl sulphates, sulphonates, phosphates and phosphonates. The products are interesting materials because the packing requirements of the large counter-ions impose some local structural ordering on the polypyrrole chains [5]. This idea can be extended to polymeric counter-ions and polymers with counter-ions derived from poly(vinyl sulfate) [127] and from poly(p-styrenesulfonate) [128] have been described. One perceived advantage in these cases is that the counter-ion is immobilized.

Polymerization of pyrrole on a platinum electrode in water requires an applied potential of at least 0.6 V v SCE. Okano et al. [129] showed that this potential is lower on an illuminated n-TiO$_2$ surface and that simultaneous electrolysis and irradiation through a mask could be used to generate conducting patterns with a line width of 45 μm.

2.3.2 Substituted Pyrroles

The effect of *N*-substitution on the polymerization of pyrrole derivatives has been studied by Diaz et al. [130]. Using cyclic voltammetry, they found that the anodic peak current for polymerization of pyrrole on a Pt electrode was at a potential of 1.2 V v SCE and the oxidation potential of the polymer was about −0.2 V. *N*-alkyl substitution had little effect on the peak potential of the monomers but increased the oxidation potential for the polymers to around 0.5 V. *N*-aryl substitution has a similar effect on the polymer but pushes the peak potential of the monomer up to about 1.8 V. *N*-substituted pyrroles are polymerizable to films, although the difficulty of producing thick films of good quality increases with the size of the substituent. Although the polymers are oxidized to comparable extents and have similar densities, their conductivities are all 5 to 6 orders of magnitude lower than the unsubstituted polymer. Street et al. [131] suggest that the loss in conductivity arises from the fact that the substituent prevents coplanarity of the rings and disrupts the effective conjugation. Although methyl substitution on the nitrogen atom reduces the conductivity by 5 orders of magnitude, substitution in the 3 and 3,4-positions reduces the conductivity by a factor of 10 only. However, this pattern of substitution also reduces the structural disorder, leading to films with higher crystallinity [132].

2.3.3 Polythiopene and its Derivatives

The electrochemical polymerization of thiophene is apparently rather similar to that of pyrrole and studies have been reported by Tourillon and Garnier [133] and by Kaneto et al. [134, 135]. Early studies are reviewed by Tourillon [136]. The oxidation potential of the monomer is significantly higher (1.6 V v SCE) than that of pyrrole and it might be expected that the more reactive cations would lead to greater structural irregularity in the polymer, which appears to be the case.

Polythiophene is usually amorphous and there is evidence from XPS and ir spectroscopy that it has significant concentrations of units linked through the 3-positions. Substitution of a single 3-position leads to a marked increase in regularity and poly(3-methylthiophene) is a more regular polymer with greater stability to electrochemical cycling [137] and a conductivity typically two orders of magnitude higher for the same counter-ion [138, 139]. Roncali et al. [140] found that the conductivity of poly(3-methylthiophene) films falls with increasing monomer concentration and suggested that high concentrations favour the formation of links to the unsubstituted 3-position of the monomer. The solubility of 3-methylthiophene in water is very low, but high enough to allow electrochemical polymerization from aqueous solution [141]. Osawa et al. [142] showed that the electrical and mechanical properties of polythiophene can be improved by ultrasonic irradiation during polymerization; they suggest that agitation reduces the level of diffusion control of the reaction at the electrode surface.

An alternative way to get better structural regularity might be to begin with bithiophene or terthiophene where some of the inter-ring bonds in the polymer are formed before the polymerization and these monomers have lower oxidation potentials (thiophene 1.6 V, bithiophene 1.2 V and terthiophene 1.0 V v SCE). Polymerizations of both bithiophene [143] and terthiophene [144] have been described but there is some doubt about whether the polymers derived from oligomers are more regular or not [145, 146]. Krische et al. [147] reported synthesis of polymers derived from the range of dimethyl-2,2'-bithiophenes; they found that the monomer with methyl groups in the two α-positions does not polymerize.

Thin films of polythiophene deposited electrochemically on a transparent electrode can be switched very rapidly from highly doped to less doped states, with a dramatic change in absorbance and they have been investigated for use in optical displays [148]. One attraction of substituted polymers is the possibility of altering the electronic structure of the polymer to produce displays with different colours. Tanaka et al. [149] have described stable conductive films of polymers and copolymers of 3-methoxy-thiophene and of 3-methoxy-2,2'-bithiophene and other reported substituents include alkyl [150], phenyl [151] and benzyl [152]. It has recently become apparent that the 3-alkyl thiophenes with long alkyl groups can be soluble in organic solvents and this development is reviewed in Sect. 2.5. Lemaire et al. [153] electropolymerized thiophenes substituted in the 3-position with chiral groups and showed that the polymers can selectively recognise chiral counter-ions introduced during doping.

2.3.4 Polyphenylene and Related Polymers

The oxidation potential of benzene is too high to allow it to be polymerized in conventional solvents; either the required potential will lead to solvent breakdown or the very energetic cations will react with the solvent. The attractions of the electrochemical route to polymer films have led to efforts to overcome this problem in a number of ways. Fauvarque et al. [154] found that it is possible to electropolymerize 1,4-dibromobenzene by electrochemical reduction of nickel phosphine complexes to yield Ni(0) species which catalyse the coupling of the bromobenzene. Using a mercury electrode, the polymer was obtained as a pale yellow powder and ir and elemental analysis indicated a chain length of only 9 rings; 10 to 15% of the weight of the product was volatile and shown to be a mixture of p-linked oligophenylenes. This procedure has since been improved to yield regular p-linked chains with lengths apparently in the range 6 to 18 units [155]; in appropriate conditions it is possible to lay down thin films of poly(p-phenylene) in its neutral state [156].

Osa et al. [157] reported as long ago as 1969 that the electrochemical oxidation of benzene in very clean conditions gave rise to oligomeric products. Several early papers described the anodic electropolymerization of benzene in unpleasant solvents. In SO_2/Me_4NBF_4 mixtures and in HF/SbF_5, the polymer grows as dendrites on the electrode [158, 159]. Dietrich et al. [160] recently reinvestigated polymerizations in SO_2 and showed that good films can be obtained with PF_6^- as the counter-ion if extreme precautions are taken in the purification of the SO_2 (normal drying with H_2SO_4 or P_2O_5 produces unidentified nucleophilic impurities). A continuous film can be produced in a two phase system consisting of 93% aqueous HF and benzene [161, 162]. Films produced in this way are reported to be free-standing but brittle and amorphous. As produced, they are insulating and, even when doped with AsF_5 from the gas phase, their conductivities are only 10^{-3} S cm^{-1}. Ir spectroscopy shows detectable amounts of o- and m-links.

Kaeriyama et al. [163] obtained insulating films of polyphenylene by electrochemical polymerization in nitromethane solution containing $AlCl_3$. Again, the films could be doped only to rather low conductivities and there was evidence of considerable structural irregularity. The first reports of a highly conducting film appear to be those of Satoh et al. [164, 165], who used nitrobenzene as the solvent and a mixture of $CuCl_2$ and $LiAsF_6$ as the electrolyte. They obtained smooth, non-passivating films with an as-grown conductivity of $100 \, S \, cm^{-1}$. The films could be undoped and redoped chemically or electrochemically and ir measurements showed no detectable defects in the p-linked structure. This route is limited by the low solubility of $CuCl_2$ in nitrobenzene and Ohsawa et al. [166] reported improved polymerization if the $CuCl_2$ was replaced by Ni or Co complexes with solubilizing ligands. Somewhat similar results were obtained by Ohsawa et al. [167, 168] using $BF_3O(C_2H_5)_2$ as the electrolyte in nitrobenzene solution. In this case there is evidence from electron diffraction that the film is crystalline and oriented, although ir and elemental analysis suggested a very defective structure.

Higher homologues of benzene have also been polymerized electrochemically. Satoh et al. [169, 170] report polymerization of naphthalene in a nitrobenzene solution containing $LiAsF_6$ and $CuCl_2$. The polymer was formed as a flexible oxidized film capable of repeated cycling between doped and undoped states. The conductivity of the fully oxidized material was much lower than that of polyphenylene, around $10^{-4} \, S \, cm^{-1}$. Similarly it was possible to polymerize anthracene to a conducting film using o-dichlorobenzene as the solvent, with $CuCl_2$ and tetrabutylammonium perchlorate as the electrolytes [171]. The product was slightly more conducting than the polymer of naphthalene. No structural studies on either polymer have yet been reported. Waltman et al. [172] reported electrochemical polymerizations of several polycyclic aromatic hydrocarbons, including pyrene and fluorene, using acetonitrile as solvent with tetraethylammonium fluoroborate as electrolyte.

2.3.5 Polyaniline

Like many other conducting polymers, polyaniline was known long before the current interest. This was in the form of 'aniline blacks', undesirable black deposits formed on the anode in electrolyses involving aniline [173, 174]. In most conditions the polymeric films are passivating but electrolysis in acidic aqueous solutions gives conducting films [175, 176, 177]. According to Ohsaka et al. [178] only in aqueous acid is the polymer formed by head-to-tail coupling; in other media head-to-head coupling gives non-conjugated polymers.

Recent studies of polyaniline have been reviewed by McDiarmid et al. [179]. They suggest that the polymer can exist in a wide range of structures, which can be regarded as copolymers of reduced (amine) and oxidized (imine) units of the form:

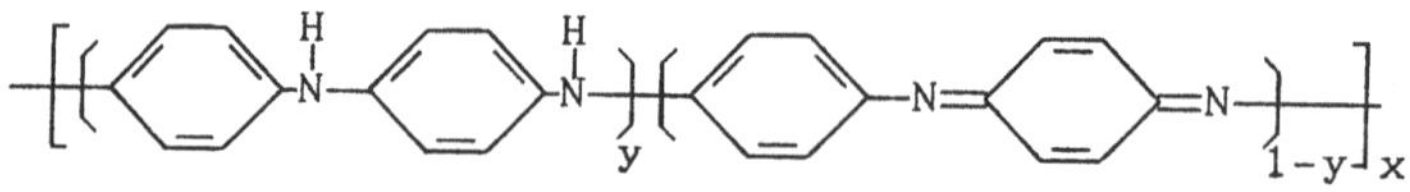

When $0 < y < 1$ these structures are the poly(p-phenyleneamineimines), in which the oxidation state of the polymer increases with increasing content of the imine form. The fully reduced form $(y = 1)$ is 'leucoemeraldine', the fully oxidized form $(y = 0)$ is 'pernigraniline', and the 50% oxidized structure $(y = 0.5)$ is 'emeraldine'. Each structure can exist as the base or as its salt, formed by protonation, so that we can envisage four repeat units in the polymer chain, in amounts which depend on the extent of both oxidation and protonation of the structure (Fig. 5).

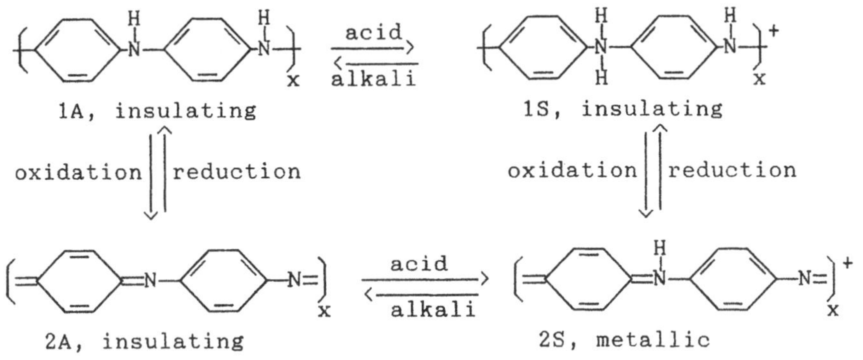

Fig. 5. The possible structures of a polyaniline chain

Wnek [180] proposed that the structure of the oxidized insulating form of conventionally formed polyaniline is approximately a 50% copolymer of diamine and diimine units, corresponding to the emeraldine structure and Hjertberg et al. [181] obtained CPMAS NMR evidence for this conclusion. Some confirmation of the structure has also been obtained by chemical synthesis of the polymer [182]. However, Kitani et al. [183] have suggested that the normal electrochemical synthesis leads to partially cross-linked polymers.

Of the various possible structures, the neutral aromatic amine and the neutral quinonediimine are both insulators and only repeat units which are both oxidized and protonated give rise to conducting segments. One consequence is that polyaniline exhibits an unusual form of doping behaviour, in that it can be switched from conducting to insulating forms by variation in pH, without any change in the electronic oxidation state [184, 185], although the exact mechanism of this response remains uncertain. The dependence of conductivity on both oxidation and protonation has been studied by MacManus et al. [186, 187]. It can be expressed in the form of a three-dimensional 'map' and results in a unique three-state switching, in which the insulating polymer can be made to switch to conducting and back to insulating with increased electrochemical oxidation. The possibility of generating conducting structures by proton abstraction from CH_2 groups in otherwise conjugated chains has been demonstrated by Hancock et al. [188, 189] but requires very strong base (nBuLi).

2.3.6 Copolymers by Electrochemical Reactions

In conventional polymer synthesis copolymerization is a common strategy for modifying polymer properties. In electrochemical polymerizations of the kind used to make conducting structures, it is expected to be difficult to make good copolymers unless the oxidation potentials of the two monomers are sufficiently close that one is not significantly preferred over the other [190].

Electrochemical polymerization of phenols to non-conducting poly(phenylene ethers) has been described [112] and Oyama et al. recently claimed that phenol [191] and p,p'-biphenol [192] can be polymerized to uncharacterized conducting films. Kumar et al. [193] claimed that it is possible to copolymerize phenol with pyrrole but not with thiophene; their paper contains no real evidence that the product is a copolymer and the small differences in the properties of the product compared to polypyrrole might well result from the effects of phenol on the nucleophilicity of the reaction medium.

Because their oxidation potentials are similar, substituted pyrroles can be copolymerized with pyrrole, allowing the limiting conductivity of the fully-doped polymer to be varied [194, 195]. The oxidation potentials of the monomers, and hence their reactivity ratios, are sensitive to the substituent [196]. Inganas et al. [197] reported the synthesis of pyrrole-thiophene copolymers starting from terthiophene, whose oxidation potential is similar to that of pyrrole. Sundaresan et al. [198] copolymerized pyrrole with 3-(pyrrol-1-yl)propanesulphonate to give a polymer in which the sulphonate counter-ion is a part of the polymer structure.

Because of the problems associated with copolymerizations of monomers of very different reactivity, many authors have looked at an alternative approach which is to synthesise appropriate sections of the desired polymer chain, then couple them electrochemically to get the final polymer. Naitoh et al. [199] synthesised the dimer, 2,2'-thienylpyrrole and used this as a monomer to prepare the alternating pyrrole-thiophene copolymer. They claimed that the copolymer film obtained with HSO_4^- as the counter-ion is more conductive than either of the corresponding homopolymers by a factor of 10 to 20. McLeod et al. [200] synthesised 2,5-dithienylpyrrole and polymerized it electrochemically with silver p-toluenesulphonate as the electrolyte. They obtained films of polymer whose conductivity could be varied in the range 10^{-8} to 0.1 S cm^{-1}. Surprisingly, some low conductivity films were soluble in acetone or acetonitrile and evaporation of the solvent gave a powder of similar conductivity. Based on the shift in the absorption maximum in the visible spectrum on polymerization, it was concluded that the soluble films were polymers with molecular weights of 4000.

Ferraris and Skiles [201] have also described a range of thiophene copolymers synthesised in a similar way.

Danielli et al. [202] prepared the alternating copolymer of benzene and thiophene by polymerization of 1,4-di-(2-thienyl)benzene and Mitsuhara et al. [203] prepared the equivalent polymers with both benzene and biphenyl units. They pointed out that cycling the polymer films between the doped and undoped states gives a green/pale yellow colour change; since conventional polythiophene films allow blue and red colours all three primary colours can be generated from conducting polymers with obvious implications for display devices. Copolymers of thiophene with both benzene and pyridine have been reported by Tanaka et al. [204]. Copolymers containing thiophene or pyrrole units, together with vinylene units, have been reported by Berlin et al. [205, 206] and similar copolymers with short chain polyene segments are reported by Tanaka et al. [207]. Jen et al. [208] reported poly(3-alkoxy-2,5-thienylene vinylenes) prepared by chemical means.

2.4 Precursor Routes to Conjugated Structures

The fibrillar morphology of Shirakawa polyacetylene is an advantage in applications requiring a high surface area but a problem in many other cases, especially the study of diffusion and transport processes and the possible device applications where reproducible junctions are required. For this reason it is desirable to be able to prepare coherent films. Although other conducting polymers, such as polypyrrole and poly-thiophene, are formed as films, their intractability is still a problem since it is difficult simultaneously to control film thickness, morphology and other properties. A standard response of the polymer chemist when faced with this sort of problem has been to use a precursor approach, in which an initially processable polymer is chemically reacted to convert it into the required polymer after it has been formed into the film or fibre. This approach has been highly successful in handling thermally stable polymer structures although the need to remove reaction products in the transformation stage does limit it to thin films and fibres. Not surprisingly, this approach has also been used in attempts to produce conducting polymers, and some of the methods are described below.

2.4.1 Polyacetylenes from Dehydrohalogenation and Related Reactions

In principle, polyacetylene should be readily formed from elimination reactions on simple polymers. Thus, the dehydrochlorination of poly(vinyl chloride) [209] leads to polyene sequences, as do the eliminations of water from poly(vinyl alcohol) [210] and of acetic acid from poly(vinyl acetate). Of these, the removal of HCl from PVC is the most facile and can be induced thermally or catalysed by a wide variety of bases [211]. Although near complete dehydrochlorination can be achieved in the thermal dehydrochlorination of PVC, the products are normally insulators and do not respond to doping. This is probably partly due to the fact that the elimination reaction is accompanied by cross-linking of the polymer which introduces sp^3 carbon atoms and limits the achievable conjugation length [212]. Another complication is likely to be the high reactivity of the polyene towards HCl, and it is not surprising that side-reactions occur. The same is true of the other, less facile thermal elimination reactions, none of which has been reported to yield a conducting polymer, presumably for the same reasons. Catalytically induced reactions of polymers are nearly always complicated by the difficulty of keeping the system sufficiently homogeneous to allow complete reaction. Even when a suitable co-solvent is available for the starting polymer and the catalyst, it is typical for the product polymer to precipitate before complete reaction.

Tsuchida et al. [213] showed that PVC solutions in tetrahydrofuran dehydrochlorinate when treated with sodium amide in liquid ammonia; the product had many of the characteristics of polyacetylene and a conductivity of 10^{-4} S cm^{-1} but was an infusible black powder. It was found that reaction of PVC powder gave only limited dehydrochlorination of the surface.

Soga et al. [214] also found poor conductivity in PVC after dehydrochlorination with base in polar solvents. Kise et al. [215] achieved some success by using a phase-transfer catalyst to induce reaction of a PVC film with aqueous NaOH in the presence of quaternary ammonium or phosphonium bromides which solubilize the inorganic catalyst into the organic polymer. They claimed that greater than 90% dehydro-chlorination could be obtained for films of thickness 20–50 μm.

The films could be doped with I_2, $FeCl_3$ or BF_3 to conductivities of the order of 10^{-3} S cm^{-1}, much lower than polyacetylene. Spectroscopy suggested that the average length of the polyene sequence was around 10 C=C units and the authors suggest that this is due to the random initiation of an unzipping reaction at many points on the chain; termination by collision of two unzipping fronts propagating in opposite directions would give sp³ C atoms. Support for this view comes from the kinetic studies reported by Howang et al. [216]. The products of PVC dehydrochlorination are air-sensitive, degrading over a period of weeks at room temperature [217]. Decker [218] has reported "writing" conductive tracks in PVC by laser photolysis but it appears that the degradation is extensive leading to graphite-like tracks.

Other variations of this theme have been described. Yie-Shun et al. [219] reported that poly(vinylidene chloride) can be dehydrochlorinated with KF and KOH complexes with crown ethers. Initial removal of one Cl atom per repeat unit is rapid but further reaction leads to aromatization as triple bonds are formed and undergo intramolecular cyclizations. The electrical properties of the films were not reported. Kise and Ogata [220] used phase-transfer catalysis by quaternary ammonium or phosphonium bromides to dehydrofluorinate poly(vinylidene fluoride) (PVDF). For powder samples the conversions reached were typically around 60–65%, corresponding to a mixture of double and triple bonds. The powder was stable in air over long periods. Reaction of thin films was limited to surface layers, the surface conductivity increasing from 4×10^{-11} to 9×10^{-8} S cm^{-1}. Doping with iodine increased conductivity to a modest 1×10^{-5} S cm^{-1}. Faingold-Murshak and Karvaly [221] suggested that dehydrofluorination of PVDF with the previous catalyst systems is very inhomogeneous, because of the poor wetting of the polymer by the aqueous base, and that it is accompanied by chain scission. They proposed the use of butyl ammonium hydroxide in *tert*-butanol solution as a milder and more controlled system which can produce thin layers of conjugated sequences on the surfaces of PVDF films and fibres; no electrical studies have yet been reported. Dias and McCarthy [222] reacted PVDF with bases in dimenthylformamide solution. They obtained conversions up to 100% to soluble polymer gels which could be cast into film form, although the resulting films were completely insoluble. Conductivities before doping were around 10^{-9} S cm^{-1}; on treatment with I$_2$ the best values obtained were 4×10^{-4} S cm^{-1}, although in several cases there was no doping effect. Iqbal et al. [223] showed that reduction of films of polytetrafluoroethylene with benzoin dianion in dimethylsulphoxide solution gives a gold coloured surface layer whose composition was shown to be a mixture of *trans*-polyacetylene and a carbonaceous fraction composed of aromatic rings and acetylenic bonds. Miyata et al. [224] showed that polymerization of 2-chloroacrylonitrile by inclusion polymerization in a host-guest complex leads to a regular polymer which can be dehydrochlorinated to a conjugated structure. Tsutsumi et al. [225] made polymers from 1-halobutadienes by the same route and also demonstrated dehydrochlorination to a conjugated structure. In both cases the conductivities were disappointing.

Another precursor route which might in principle be used to produce polyacetylene is the dehydrogenation of polybutadiene. Unfortunately, it has not so far been possible to achieve conversions high enough to allow this to be a feasible route — precipitation of the polyene from the reaction mixture limits the conjugation length. The same is true, but more so, of attempts to prepare conjugated polyenes from polyisoprene, polystyrene and poly(1,3-cyclohexadiene) [42]. Schomaker and Tolbert [226] synthesised *n*-doped polyacetylene directly by the deprotonation of a copolymer of acetylene and 1,3-butadiene. The starting copolymer is soluble in THF but of relatively low molecular weight (<200 units); the deprotonated product has only very short conjugated sequences. It seems fair to say that preparation of con-jugated polyenes by elimination reactions of simple polymers is of limited use because of the complexities introduced by side-reactions. The products certainly do not meet the requirements of the physicist for characterised structures!

2.4.2 The Durham Route to Polyacetylene

The Durham precursor route to polyacetylene is an excellent example of the applica-tion of organic synthesis to produce a precursor polymer whose structure is designed for facile conversion to polyacetylene. Durham polyacetylene was first disclosed by Edwards and Feast, working at the University of Durham, in 1980 [227]. The polymer (Fig. 6 (I)) is effectively the Diels-Alder adduct of an aromatic residue across alternate double bonds of polyacetylene. The Diels-Alder reaction is not feasible, partly for thermodynamic reasons and partly because it would require the polymer to be in the all *cis*-conformation to give the required geometry for the addition to take placed [228]. However, the polymer can be synthesised by metathesis polymerization of the appropriate monomer.

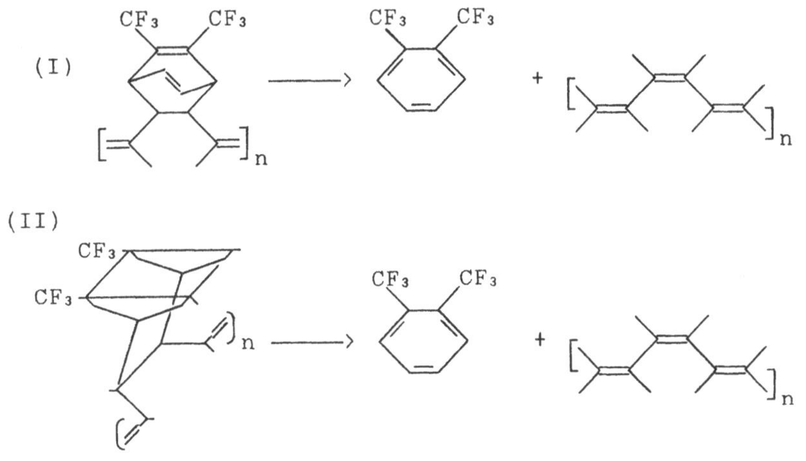

Fig. 6. "Durham" precursor routes to polyacetylene

The aromatic residue may be any of a large number of such units but the favourite for academic study has been the perfluoromethylxylene derivative shown, which smoothly eliminates at around room temperature to give a polyacetylene containing 25% of *trans*- and 75% of *cis*-units. After transformation and isomerization at 80 °C, the polyacetylene produced is a continuous dense film. The physical chemistry of the transformation and isomerization reactions has been studied in detail [229, 230] and the properties of the polyacetylene are reviewed [231]. The great advantage of this route is that the precursor is a soluble polymer so that it can be characterized and the physical form of the polyacetylene can be controlled.

The perfluoromethylxylene group eliminated in the transformation of the precursor is very convenient because it is volatile enough to be easily removed from the polyacetylene. However, the retro-Diels-Alder reaction is rather too rapid at room temperature and the precursor has to be stored at low temperature. Feast and Winter [232] have used metathesis polymerization of the isomer (II) to produce a precursor in which the elimination still yields perfluoromethylxylene but is now symmetry-forbidden and much slower. This polymer can be stored and handled at room temperature but its elimination reaction is rather exothermic.

2.4.3 Polymers via Sulfonium Eliminations

Another route to a conjugated polymer structure via a soluble precursor, which has been extensively studied, involves the preparation of poly(p-phenylenevinylene), in which the chain consists of alternating aromatic rings and carbon-carbon double bonds. Wnek et al. [233] prepared the polymer by the Wittig reaction, as described earlier. This method gave insoluble oligomeric powders with conductivities up to 3 S cm^{-1} at most. A very large range of derivatives of poly(p-phenylene vinylene) has been chemically synthesised and their preparation and properties have been reviewed by Horhold and Helbig [234]. Phenyl-substituted phenylene vinylene polymers are soluble in organic solvents and can be cast into films which are dopable to semiconductor levels.

A more convenient route to poly(p-phenylene vinylene) is via a soluble sulfonium salt precursor:

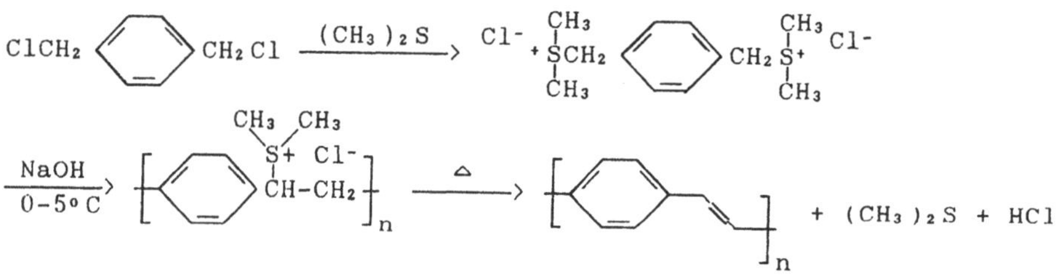

This route was described more or less simultaneously by Karasz et al. [235] and by Murase et al. [236]. The precursor polymer is typically prepared by reaction of the appropriate bis(chloromethyl)arylene compound with dimethylsulfide in a polar solvent [237, 238]. The product is a water-soluble polymer which can be cast to give thin films. Elimination of hydrochloric acid and dimethyl sulfide takes place on heating the film in the range 200 to 300 °C and can be monitored by thermogravimetry and by the development of colour and conductivity [239]. Poly(p-phenylene vinylene), prepared by the precursor route, can be doped to much higher conductivities than the conventionally synthesised polymer.

The sulfonium elimination route can easily be extended to prepare substituted polymers and Murase et al. [240] showed that polymers with alkoxy substituents in the phenyl rings can be doped more easily and to higher conductivities than can poly(p-phenylene vinylene) itself. Karasz et al. [241] showed that substitution of the phenyl rings with groups of differing electronegativity changes the electron density on the polymer backbone and alters reactivity with dopants and the stability of the doped polymer. In contrast, substitution with groups of constant electronegativity but differing size changes the chain packing and alters the conductivity and processability of the polymers. Karasz et al. [242] have also prepared copolymers w,th substituted and unsubstituted phenylene vinylene units and found that the introduction of relatively small concentrations of alkoxy-substituted units allows easier doping of the copolymer. Antoun et al. [243] have used the same chemistry to prepare poly(1,4-naphthalene vinylene).

Several authors have recently adapted the sulfonium elimination route to prepare poly(2,5-thienylenevinylene) and reported high conductivities (200 S cm^{-1}) on doping [244, 245]. In this case the sulfonium precursor is insoluble in water but can be cast from dimethylformamide solution. Jen et al. [246] report an alternative route involving a water-soluble precursor which eliminates HCl and tetrahydrothiophene.

2.4.4 Precursor Routes to Poly(p-Phenylene)

Poly(p-Phenylene) is probably the most thermally and oxidatively stable conducting polymer known and has been of interest to polymer scientists both for its conducting properties and for its stability. Its synthesis by step-reaction polymerization or by chemical or electrochemical oxidation invariably gives powders or rather poor quality films, so that a precursor route would be very attractive.

Marvel et al. [247] attempted synthesis of poly(p-phenylene) from poly(1,3-cyclo-hexadiene), produced by Ziegler initiation:

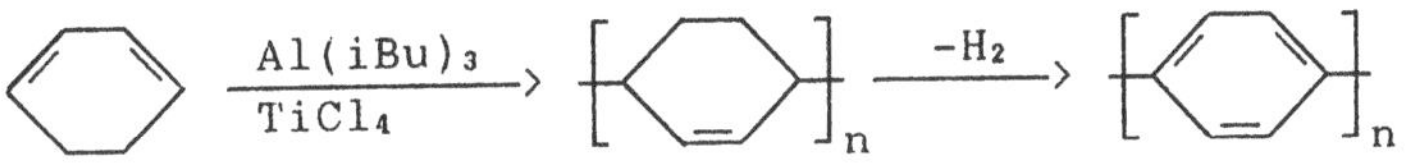

Aromatization of the polycyclohexadiene by dehydrogenation with a range of reagents, including palladium, chloranil and N-bromosuccinimde, was only partially successful, partly because both the precursor and the product polymer are insoluble. Attempts to improve the aromatization by reaction of the polycyclohexadiene with Br$_2$, followed by thermal elimination of HBr were also not very successful [248, 249] and there is the additional complication that the HBr produced may react with the polyene.

Ballard et al. [250] have described synthesis of poly(p-phenylene) via the polymerization of esters of 5,6-cis-dihydroxycyclohexadiene. The polymer, produced by free-radical initiation, is soluble in common solvents and smoothly eliminates in the temperature range 140 to 240 °C to give the aromatic polymer.

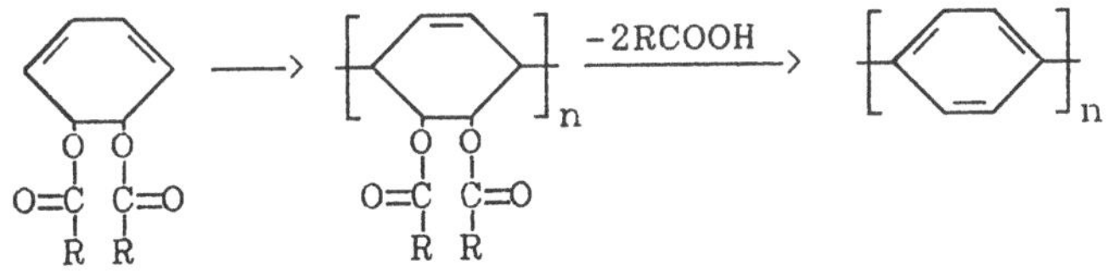

Conventional synthetic chemistry needs a multi-step synthesis to give the cyclo-hexadiene diol in the trans-form, which is difficult to polymerize [251]. A most ingenious feature of the work of Ballard et al. was the use of a bacterial fermentation route to prepare the diol in the cis-form. A detailed account of the preparation, properties and transformation of the polymer has recently been published [252]. In practice, the favoured esters are those of methylcarbonic acid, since the eliminated acid decomposes at transformation temperatures to yield methanol and CO$_2$. The main problem with this route is that the free-radical polymerization of dienes is not wholly regiospecific and the precursor polymer has about 15% of units resulting from 1,2-addition reactions, leading to a polyphenylene which is not completely p-linked. Nevertheless, the product can be doped with AsF_5 to conductivities of the order of 100 S cm^{-1}. It is also of interest as a high-temperature, stable, coating material and as a dielectric.

2.5 Soluble Polymers and Latices

Polyacetylene prepared by the Shirakawa route pyrolyses on heating, before showing any detectable crystal melting point. At the same time, it is insoluble in all known solvents. For these reasons it is essentially unprocessable. Until recently it has seemed to be a general rule that all conducting polymers were insoluble, which follows natur-ally from the conjugation of the double bonds along the chain which results in chain stiffness.

Many authors have looked for ways of making more tractable polymers. One main method has been the production of block and graft copolymers with conventional polymers and this is discussed in Sect. 2.6. Another approach has been to try to form the polymers as colloidal dispersions from which films can be formed. Edwards et al. [253] reported the synthesis of polyacetylene latices having particle sizes of the order of 40 to 200 nm diameter. The synthesis was achieved using a Luttinger catalyst in a solution containing a poly(t-butyl styrene) — poly(ethylene oxide) block co-polymer. Little information on the properties of the polymer is given except that the particles could be recovered and pressed into pellets with conductivities comparable to conventional *trans*-polyacetylene films. Kminek and Trekoval [14] found that Shira-kawa polyacetylene can be disintegrated ultrasonically to dispersions of particles with a hydrodynamic radius of around 300 nm, from which films could be recovered by evaporation of the solvent. The films have a surface area of $>300 \text{ m}^2 \text{ g}^{-1}$, about 5 times greater than the original polyacetylene and poor mechanical coherence. Spongy films could also be deposited electrolytically if prepared with an ionic surfactant. Armes and Vincent [254] have prepared aqueous latices of polypyrrole and Yassar et al. [255] used chemical polymerization of pyrrole to prepare submicron latices of sulfonated or carboxylated polystyrene coated with polypyrrole.

In 1982, Soga et al. [256] showed that exposure of acetylene to AsF_5 at low temperatu-res leads to rapid polymerization (in our experience this reaction can be explosively violent). The product is a solid polymer which is heavily arsenic-doped and has a conductivity several orders of magnitude lower than a conventional sample of poly-acetylene saturation-doped from the gas phase. Aldissi and Liepins [257] have adapted this reaction to the preparation of soluble polyacetylene by adopting AsF_3 as the reaction solvent. They claim that polymerization of acetylene with AsF_5 is very rapid, giving a polymer which is soluble in common solvents. However, elemental analysis shows that the polyacetylene formed contains about one As atom per 10 CH units and this is not removed on repeated reprecipitations. It seems likely that the As atoms form part of the chain backbone, conferring sufficient flexibility to allow dissolution. It is claimed that films of soluble polyacetylene can be doped but very little information has been published.

The use of AsF_3/AsF_5 mixtures as solvents for heavily doped conducting polymers and as simultaneous solvent, dopant and polymerization initiator has been elaborated by workers at Allied Chemical Co. and reviewed by Frommer [258]. Many of the common conjugated polymers have been prepared in solution in this aggressive mixture and it is suggested that solvation arises mainly from the interaction of the solvent with the dopant counter-ions [259].

During 1986, the search for soluble conducting polymers was greatly advanced when groups at Tsukuba [260] and at Allied Chemical [261] reported that neutral poly-(alkylthiophenes) are soluble in common organic solvents when the β-positions are substituted with alkyl groups which are butyl or longer. These polythiophenes have been prepared electrochemically, by iron (III) chloride oxidation [101] and by Grignard coupling of di-iodothiophenes, and become soluble when undoped. Poly(3-hexylthiophene) has been doped to form stable blue solutions using $NOPF_6$ [262]. In other cases the soluble polythiophenes are only soluble in the undoped state.

The availability of soluble polymers has permitted measurements of molecular weight, as discussed below. The glass transition temperatures are quite low, from 145 °C for polymethylthiophene to 41 °C for polybutylthiophene. X-ray diffraction shows a broad peak which has been interpreted by different groups as showing a structure which is either partly crystalline or is amorphous. The temperature of fusion decreases from 280 °C to 80 °C as the length of the alkyl chain increases from 4 to 22 carbons [263].

When doped, these polymers show conductivities in the range from 0.5 to 50 S cm^{-1}, as illustrated in Table 1 (page 17) Kaeriyama and co-workers [264] show from optical spectroscopy that the band gaps increase with the size of the side group. In the doped states the spectra were all similar, but the conductivity drops with the size of the side group.

Closer reading of the published work, reveals that the polymers are not wholly soluble and that the extent of solubility is dependent on the solvent [265]. In the case of a butylthiophene-methylthiophene copolymer, hot THF solubilizes a fraction of low molecular weight which forms brittle films, while hot xylene extracts a higher molecular weight polymer which forms tough films. There remains an insoluble, and presumably cross-linked residue. The microstructure of the polymer is presumably quite dependent on the polymerization method and conditions.

While the polymers seem to be amorphous in the solid state, there are signs of a transition in solution, possibly from a disordered coil state to a helix, which involves a temperature-dependent colour change [266]. This change is also associated with precipitation of a partly-crystalline form. The crystals and solution apparently co-exist over a rather wide temperature range, which is unusual because polymers generally show a rather sharp transition from solubility to insolubility. This may reflect a broad distribution of molecular weights in the polythiophenes.

The range of soluble polythiophenes has been growing rapidly to include side chains up to docosane [267], ether and amide links in the side chains [268], and water-soluble polymers with sulfonated side chains (Table 1) which are claimed to be "self-doping" in that the sulphonate may act as the counterion to the delocalized chain cation [269, 270]. In principle, these polymers can be p-doped and undoped by the transport of a proton or a small cation rather than a large anion, and so may respond more rapidly. By treatment of an aqueous solution with $NOPF_6$, a doped solution can be made, which slowly degrades.

The fusion temperature of these polymers is low enough to allow the spinning of fibres and melt pressing of films [263]. They can also be blended with normal thermoplastics such as polystyrene or poly(ethylene oxide) [271]. The conductivity shows a percolation threshold of about 16% which is expected for a random distribution of conducting spheres.

Soluble polyaniline has been reported as being prepared in the presence of aromatic sulphonic acids [272].

2.6 Blends, Block- and Graft-Copolymers, and Composites

In order to prepare useful materials from conducting polymers, there is a need for good mechanical properties and environmental stability as well as conductivity. An obvious way to produce these polymers in a tough, processable, stable form is to make composites with stable, passive polymers. In principle, this can be achieved by block or graft copolymerization, by synthesis of the conducting polymer within the host, or by simple blending. The flow properties and mechanical properties of the blend should be dominated by the matrix polymer at low levels of conductor but the composite will become increasingly stiff as the conducting-polymer phase becomes continuous rather than isolated particles (Fig. 7). This change in properties will depend on whether the conducting polymer is present as fibres or as equiaxed particles. The conduction properties of the composite will be governed by the need for percolation of the charge carriers between particles, with the conductivity rising rapidly at the point at which the conductor forms a continuous network, usually about 20% by volume for spherical particles [273] (Fig. 8).

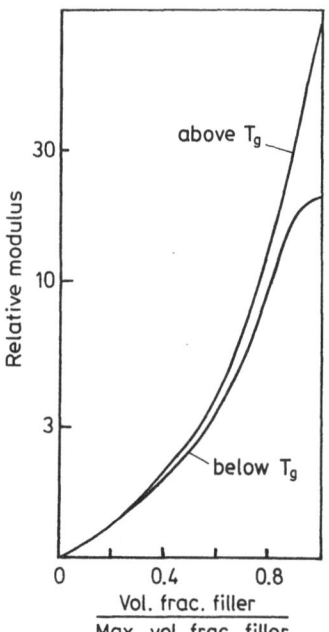

Fig. 7. Dependence of modulus on filler content for a polymer containing a hard, particulate filler. Reproduced with permission from Ferringo TH (1987) In: Katz HS, Milewski JV (eds) Handbook of fillers for plastics, Van Nostrand Reinhold

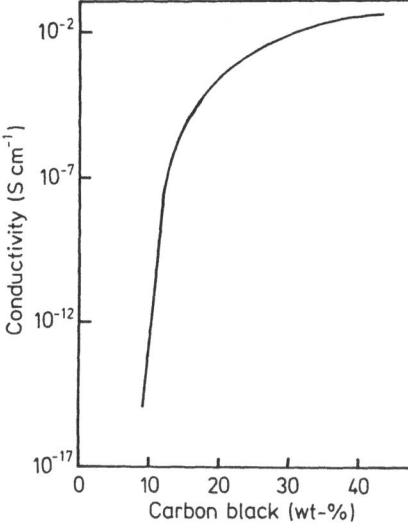

Fig. 8. Dependence of conductivity on filler content for a polymer containing carbon black, a particulate conducting filler. Reproduced with permission from Blythe AR (1979) Electrical properties polymers, Cambridge University Press

Polyacetylene blends were prepared by Galvin and Wnek [274] by the polymerization of acetylene in films of polyethylene doped with a Ziegler catalyst. The polyacetylene forms as a particulate second phase with a size from 60 to 200 nm. The yield point of the low-density polyethylene films increased from 7 MPa to 10 MPa with 7% polyacetylene and the extension to break was reduced. This effect was greater in blends produced by polymerization in solid polyethylene, presumably because there was a greater tendency for the polyacetylene to form a network surrounding the polyethylene crystallites. The modulus of samples produced by low-temperature polymerization was also much higher than that for the original polyethylene or the samples produced at high temperatures. The conductivity of iodine-doped material increased from very low values to about 1 S cm^{-1} at 3 wt% polyacetylene and then increased further to 10 S cm^{-1} at 25%. This low percolation threshold is somewhat surprising, even when it is corrected for the increase in volume on iodine doping, and for the 50% crystallinity of the matrix polymer. The authors attribute this to the very finely divided nature of the polyacetylene formed.

Some related studies have been made of polymerization of acetylene in polybutadiene rubber [275, 276]. The morphology and electrical properties of these blends have been reviewed [277]. In this case, the percolation threshold for conductivity on iodine doping is closer to 20 vol% of polyacetylene, at which point the polyacetylene seems to form a continuous matrix and the modulus of the rubber starts to rise rapidly. A conductivity of 10 S cm^{-1} was achieved at polyacetylene loadings above 30%. Lee and Jopson [278] polymerized acetylene in ethylene-propylene-diene rubbers and produced conducting, tough films. Stretching the films produced large increases in conductivity. Wessling and Volk [279] report mechanically blending various conducting polymers into thermoplastics to make a processable mix, though with low conductivity ($< 10^{-5}$ S cm^{-1}).

Another approach to blending of polyacetylene with tough polymers is to form graft or block copolymers [280, 281]. Aldissi [282] produced block copolymers by polymerizing acetylene at the ends of chains of anionic polyisoprene after conversion of the anionic chain ends to Ziegler centres. The polymers appeared to be soluble, with the polyacetylene segments having molecular weights of 1500 to 1900, measured by gel permeation chromatography. The electrical properties of the products have been reviewed [283, 284]; they show a percolation threshold at about 16 vol% polyacetylene, as do similar block copolymers of polystyrene and polyacetylene [285], which agrees well with theoretical estimates. Stowell et al. [286] synthesised acetylene-styrene block copolymers by an anionic to methathesis conversion of the polystyrene chain ends but do not describe the electrical properties of the product. The same authors have also described block copolymers of acetylene and cyclopentene, prepared by sequential addition of the two monomers to a living metathesis catalyst [287].

Armes et al. [288] found difficulty in repeating syntheses of isoprene-acetylene block polymers by the anionic — Ziegler route and suggested that the polyisoprenyl-titanium intermediate is very sensitive to impurities. They used reaction of anionic polyisoprene with $CoCl_2$ to make an analogue of the Luttinger catalyst, attached to the chain ends and reported successful synthesis of the block copolymer, although little detail of its properties is given.

Poly(methylmethacrylate) side chains have been grafted on to polyacetylene by treating sodium-doped polyacetylene with the MMA monomer [14]. The polymers were soluble at mole ratios of MMA to acetylene of greater than 1.5. A graft copolymer with polyacetylene side-chains on a polybutadiene backbone has also been reported as soluble [289]. Morphological studies of this material [290] show worm-like aggregates of polyacetylene, 100 to 400 nm long and 10 to 50 nm wide, embedded in the rubber. There are also polyacetylene microdomains (10 nm) in the polybutadiene matrix and polybutadiene domains in the polyacetylene aggregates, a structure very reminiscent of ABS (acrylonitrile-butadiene-styrene graft copolymer). In these films the onset of high conductivity ($>1 \text{ S cm}^{-1}$) occurs at about 50 wt% polyacetylene. Light scattering studies on these polymers give molecular weights much higher than expected for the known polyacetylene-polybutadiene ratio with a radius of gyration of 20 nm versus 6 nm for the polybutadiene alone. This shows that the polymer is aggregated in solution, possibly as a micelle containing highly ordered polyacetylene subunits [291, 292, 293]. These light-scattering studies contrast with much lower molecular weight measurements on similar systems by GPC. Spectroscopic studies of these composites have been described [294].

The electrochemical polymerization of pyrrole or thiophene readily lends itself to formation of composites. Polypyrrole-acetylene laminates have been made by using polyacetylene as an electrode [295]. The polypyrrole forms as a 5 μm skin on the polyacetylene. If the polyacetylene is first doped, the polypyrrole completely permeates the film. In both cases the conductivity of the composite reached 30–40 S cm⁻¹ and was much less sensitive than that of pure polyacetylene to exposure to moist air or water, so that the polypyrrole protects the polyacetylene even in the case where it permeates the film. In this latter case, treatment with ammonia caused the conductivity to drop by $30\times$ whereas for the sandwich films the conductivity dropped by $4600\times$ through the film but only $17\times$ in the surface layers.

Two independent groups have reported that polypyrrole can be formed within a swollen film of poly(vinyl chloride) attached to an electrode surface [296, 297]. Niwa et al. [298, 299] have described the properties of these composites, which, they claim, can combine the toughness of PVC with the conductivity of polypyrrole.

At short times of electrochemical polymerization only the electrode face of the composite is conducting; at longer times the polypyrrole grows through the film and both faces and the bulk become conducting. The process apparently depends on the ability of acetonitrile strongly to swell the PVC without actually dissolving it. The morphology of the polypyrrole phase varies through the thickness of the film and depends on the supporting polymer. By electropolymerization onto patterned electrodes, it is possible to produce composites with rods or planes of polypyrrole within PVC, and highly anisotropic conductivity [300].

A similar process has been used to make polypyrrole-poly(vinyl alcohol) composites by electropolymerization from aqueous solution into a cross-linked poly(vinyl alcohol) film [301]. The conductivity decreased and the elongation to break increased $10 \times$ with respect to polypyrrole. A conducting composite of 6% polypyrrole in a vinylidenefluoride-trifluorethylene copolymer showed an extension to break of 120% [302].

By electropolymerization of pyrrole in solvents containing polyelectrolytes such as potassium polyvinylsulfate, it is possible to prepare films of polypyrrole with polymeric counterions which have good conductivity (1–10 S cm^{-1}) and strength (49 MPa) [303, 304, 305]. Such a material could be used reversibly to absorb cations in an ion exchange system. Pyrrole has also been electrochemically polymerized in microporous polytetrafluoroethylene membranes (Gore-tex), impregnated with a perfluorosulphonate ionomer [306].

Composites can also be prepared by electropolymerization from solutions containing dissolved polymer [307]. Since films of polypyrrole or polythiophene are normally porous, it seems most likely that the dissolved polymer is simply entrained in the pores. Similarly, composites have been prepared by polymerization of pyrrole in the presence of acrylic latex, giving blends with 10–30% polypyrrole that are conducting yet processable [808]. Presumably the polypyrrole is distributed throughout the latex particles.

Polypyrrole can also be prepared by chemical oxidation of aqueous pyrrole solutions. Composites have been prepared by introducing a polymer film at the interface between aqueous FeCl$_3$ and pyrrole in toluene [309, 310]. This method allows the preparation of polypyrrole-impregnated composites of cloth and paper as well as other polymers, including ionomeric membranes [311]. A similar route is to prepare colloidal polypyrrole in aqueous polymer solutions from which films can be cast. Films of polypyrrole in methyl cellulose gave a conductivity of 0.2 S cm^{-1} [312]. Polypyrrole with incorporated porphyrins and other dyes has also been prepared [304]. Pyrrole can also be polymerized from the vapour phase [313] and Yosomiya et al. [314] adapted this route to prepare composites. They used complexes of CuCl$_2$ and of FeCl$_3$ with poly(vinyl alcohol), PVC and PMMA to initiative the polymerization and obtained transparent films of high conductivity. From conductivity studies they concluded that the conductivity is mainly dominated by thermal hopping.

Polythiophene lends itself to the same routes to composites. A poly(3-methylthiophene)-poly(methylmethacrylate) composite has been made by electrochemical polymerization from a solution of thiophene and PMMA in methylene chloride and nitrobenzene. At high current densities the electrode side quickly became highly conducting while the outer side was less so [307]. Similar composites have been prepared by chemical routes, using a Grignard reaction, firstly to couple the thiophene units in a step-reaction, then to initiate the polymerization of the methyl methacrylate [315].

An interesting variant on this theme is the polypyrrole-polythiophene *n-n* junction
di
re
of

by mixing solutions of the sulfonium precursor with polyethyleneoxide, hydroxy-
propylcellulose and polyvinylmethylether [317]. Polyethyleneoxide forms spherulites
which impose a spherulitic texture to the polyphenylenevinylene that is retained after
transformation. As a result of this open network, high conductivities are reached
at only 10% conducting polymer.

In all of these cases, it does seem that the law that "you do not get something for
nothing" applies. The conductivity decreases as would be expected from percolation
theory, if the conducting polymer is dispersed, and according to the volume fraction
if it forms a network. The mechanical properties improve as one might expect from an
additivity rule applied to the two components.

3 Microstructure of Chains

Except in rather unusual circumstances, the synthesis of polymers by coupling of
small molecules rarely proceeds as the stoichiometric equation implies. Even where
the degree of control over the reaction is high, the polymer produced is almost always
a mixture of chains of differing lengths so that consideration has to be given to determining both the average molecular weights and the distribution of chain lengths.

At its simplest, the structure of a polymer derived from a single repeat unit should
be a linear chain, free of defects. In reality, this ideal is rarely reached since most of
the conventional methods of forming polymers have side-reactions which lead to
the introduction of anomalous links, typically in the form of isomeric repeat units,
branches or cross-links. The presence of irregular structures in the chains of conducting
polymers is expected to be significant, since it will limit both the conjugation length
and the extent of crystalline order attainable. In other polymers cross-linking would
also lead to a loss of solubility and melt-flow but this is not a problem in the conducting
polymers since they are usually inherently intractable.

3.1 Molecular Weights

Polymer science is usually concerned with flexible, high molecular weight, linear
chain molecules, which are fusible and soluble. As illustrated in Fig. 9, characteristic
properties such as viscosity and toughness only appear at high molecular weights.
According to current thinking, the key feature is the appearance of entanglements
when the chains become sufficiently long. The length at which that happens depends
on the chain flexibility. From melt viscosity studies the molecular weight between
entanglements is estimated as about 3800 in polyethylene and 36,000 in polystyrene,
where the chain is more crowded and so stiffer [318].

The linear polymers can be contrasted with the relatively weak and brittle organic
glasses, including heavily cross-linked resins and molecular glasses, such as quenched

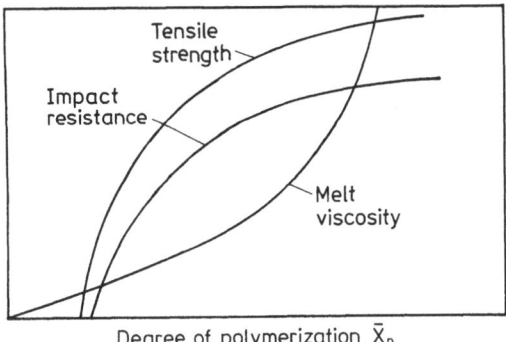

Degree of polymerization $\bar{X}_n$

Fig. 9. Effect of molecular weight on polymer properties

sugars. Organic glasses and molecular crystals may have similar elastic, optical and electronic properties to many polymers but lack the properties that arise from the long chain structure, including toughness, orientability, and film and fibre formation.

Molecular weights of linear polymers are generally determined by solution measurements [319]. Osmometry and light scattering studies can give absolute values of molecular weight averages and solution viscosity provides a simple method which must be calibrated with samples of known molecular weight. Gel-permeation chromatography will give a full molecular weight distribution if properly calibrated or used with a detector sensitive to molecular weight. In the absence of convenient solvent systems, it is common practice in industry to compare molecular weights on the basis of melt-viscosity measurements. Failing any of these techniques, it is possible to determine molecular weights by end-group analysis or by selective labelling of the chain ends. These latter methods are notoriously unreliable as they require the detection of very small concentrations of end-groups and they are vulnerable to processes which may produce chains with unexpected end groups or with multiple ends, such as chain transfer during polymerization, or chain branching.

With the exception of the alkyl-substituted thiophene polymers discussed above and the solutions of conjugated polymers in extremely aggressive solvents, such as AsF_3, virtually all of the conducting polymer systems are insoluble and infusible in the conducting state and most are also insoluble in their initial as-polymerized state. This means that most molecular weight studies must depend on end-group analysis or on converting the polymer to a soluble form with the danger of modifying the molecular weight during the process.

Shirakawa [320] hydrogenated polyacetylene which had been sodium-doped to expand the structure. The resulting 'polyethylene' was 60% soluble, the remainder being a gel. The soluble fraction had an average molecular weight of 6200. The hydrogenation conditions, 200 °C for 7 hours, are quite severe and it would not be surprising if some cross-linking and degradation occurred. Enkelmann et al. [321] chlorinated polyacetylene produced with the Luttinger catalyst and obtained a product with a molecular weight of 5900. Prolonged storage of the initial polyacetylene at −30 °C gave increasing amounts of insoluble product after chlorination.

Mechanistic studies of polymerizations initiated by Ziegler catalysts often make use of termination by radiolabelled additives, combined with molecular weight measurement by solution techniques, as a way of counting chain ends. In a conventional alkene polymerization, the polymer chains are bound either to Ti or to Al, provided that there is no significant chain transfer to monomer or β-elimination to give free chain ends. Reaction with tritiated methanol will detach all polymer chains from the metal atoms, leaving them tritium labelled, whilst reaction with ^{14}CO is more selective for polymer-titanium bonds. In principle this method can be reversed for polymerizations of acetylene on the Shirakawa catalyst and the labelling of the polymer used to count the chain ends and determine the molecular weight. A drawback of this method is that any chain transfer will lead to polyacetylene chains that are not attached to metal atoms and so do not get labelled, leading to an overestimate of molecular weight.

Chien et al. [322] made a detailed study of the kinetics of acetylene polymerization using the Shirakawa catalyst. They used both tritiated methanol and ^{14}CO to label the polymer and estimate the molecular weights and the concentration of active sites in the reaction. Whereas CO is a terminating agent for most Ziegler catalysts it does not terminate acetylene polymerization unless the monomer is simultaneously removed. Instead, it becomes incorporated into the polymer as a comonomer, allowing the possibility of producing acetylene CO copolymers [323]. On the basis of kinetic and labelling studies, it was concluded that chain transfer and β-elimination are relatively unimportant and that the labelling method, using tritiated methanol, can be used to estimate molecular weights.

Chien [6] has investigated the effect of polymerization conditions on acetylene polymerizations and has found molecular weights in the range of 5,000 to 100,000 with substantial variations between the powder and gel phases of the polymer. In normal Shirakawa polymerizations the molecular weight was said to be typically around 10^4. By controlling the conditions, Chien and Schen [324] were able to obtain polymers with molecular weights ranging from 400 to above 10^6, an enormous range for a single catalyst system. Even if these molecular weights were in error, their relative values should be reliable. It seems unlikely that there is a gross overestimate in the case of the lowest molecular weights, since much lower molecular weights would tend to give soluble polymers. It has been shown [325] that this variation in molecular weight has little effect on the conductivity and transport properties of the films, suggesting that charge transport in polyacetylene is limited by inter-chain hopping rather than by migration along the chains. Thus the overall situation for polyacetylene is rather unsatisfactory but current measurements indicate normal molecular weights to be well into the range of true polymers.

Polyacetylene may also be produced from a soluble precursor polymer by the Durham route, described earlier. In this case the soluble precursor can be studied by conventional solution methods, provided that it is kept cold enough to prevent transformation. The molecular weight of the precursor has been determined by light scattering and low-temperature GPC [326] and corresponds to a polyacetylene chain with a molecular weight of about 200,000, with M_w/M_n of about 2.

In contrast to polymers produced by chain reactions, those derived from step-reaction routes are expected to have lower molecular weights. Naarman [327] has used ir spectroscopy to estimate the phenyl end-groups, and so the molecular weights, of polyphenylenes produced by the Kovacik method. By increasing the reaction temperature from 5 to 36 °C, he was able to increase the degree of polymerization from 10 to 45. Kovacik and co-workers [328] used a similar approach and have estimated average degrees of polymerization of greater than 10 for the same system. However it must be realised that this analysis depends on very weak bands in the ir spectrum. ^{13}C NMR has demonstrated that polyphenylenes are predominantly p-linked but is not sensitive enough to detect structural defects [329, 330]. Attempts to hydrogenate polyphenylene to render it soluble require very severe conditions and are not very successful [331]. Alkylation under Friedel-Craft conditions leads to a soluble, partially alkylated product whose molecular weight corresponds to an average of 15 repeat units per chain [332].

Recently Kovacic et al. [333] have reexamined polyphenylene using Laser Desorption Fourier Transform Mass Spectroscopy. They claim that the very rapid desorption of the sample leads to a plasma containing the original molecular ions and that there is no significant fragmentation of the chains. The bulk of the molecules in samples of polymer prepared by the oxidative coupling method were found to be linear chains with varying amounts of chlorine substitution in the end rings. Samples prepared by the Yamamoto route were typically shorter than those from the Kovacic route although there was a lot of overlap. Samples prepared with the Kovacic $AlCl_3/CuCl_2$ catalyst contained fragments with mass 2 amu below the corresponding linear chains; these were attributed to cyclization reactions, leading to polynuclear structures in the chains.

The polyphenylenes prepared by step-reaction routes are thus at best short-chain oligomers. The preparation of polyphenylene from the cyclohexadiene diester, described by Ballard et al. [249, 251], involves a soluble precursor polymer whose molecular weight has been estimated by light and neutron scattering. It is reported that the polymer has a chain length of 600 to 1000 units and the precursor is shown by neutron scattering to adopt a random coil conformation in the solid state. As the precursor is isomerized to the poly(p-phenylene), the intrinsic viscosity rises rapidly, as does the apparent molecular weight as estimated by GPC. This is apparently due to the increase in hydrodynamic volume induced by putting increasing numbers of stiff segments into the flexible precursor chain. The long chain length of polyphenylenes derived from this route appears to have little effect on their limiting conductivities in the doped state, which are very similar to that of the short-chain Kovacic material. As far we are aware, there are no published data on the molecular weight of electrochemically synthesised polyphenylene. A study of the conductivity of polyphenylene oligomers after doping with potassium, showed that the conductivity was $<10^{-6}$ S cm^{-1} up to four rings (quadriphenyl), then about 10^{-2} S cm^{-1} for five and six rings, and >10 S cm^{-1} for polymer [334].

Because of the total insolubility of polypyrrole, and polythiophene, no molecular weight data are available for these polymers, although Cao et al. [335] reported studies of an oligomeric, chloroform-soluble fraction of polythiophene, obtained from a polymerization by Grignard coupling. Studies have been carried out on 3,4-dimethylpyrrole by observing the elimination of tritium from labelled 2 and 5 positions [336]. With the assumption that the residual tritium is on the chain ends only, this yielded degrees of polymerization from 100 to 1000. These are lower limits since it is possible that saturated carbon atoms are left in the chain by failure to rearomatise completely; these data show that true high polymers are formed. The availability of the soluble polymers of 3-alkyl thiophenes offers the possibility of direct molecular weight measurement, although few authors have yet taken advantage of this. Hotta et al. [337]

estimated the molecular weight of electrochemically prepared poly(3-hexyl thiophene) as around 20,000, corresponding to about 150 units. Watanabe et al. [338] reported a value of 9000 for the molecular weight of a soluble fraction extracted from an electro-polymerized polyaniline sample.

One of the few processable polymers capable of doping to respectable conductivity is poly(phenylene sulfide). Stacey [339] has reported GPC studies of this polymer, using an instrument operating at 210 °C with 1-chloronaphthalene as solvent and detection by viscosimetry. Samples tested were found to be of high molecular weight with M_w in the range 20 to 80,000 and M_w/M_n typically less than 1.7.

There is little molecular weight information available on other systems. In general it seems to be the case that polyacetylene and the precursor systems yield degrees of polymerization which are certainly within the high polymer range, while most of the step-reaction polymerizations yield only low degrees of polymerization. Electro-polymerization to soluble polymers apparently gives intermediate chain lengths.

3.2 Cross-Linking, Branching and Other Defects

The question of cross-linking in polyacetylene has been discussed in detail by Chien [6]. Studies on polymethylacetylene showed no cross-linking of this soluble polymer either on oxidation or heating in-vacuo but it might be argued that cross-linking is more likely in polyacetylene due to the closer packing of the chains. Li et al. [340] subjected polyacetylene to exhaustive ozonolysis and methylation of the ozonolysis products. They found only the expected product, dimethyl oxalate, with no detectable products arising from cross-links. ^{13}C NMR of the solid polymer showed no sp^3 carbons in Shirakawa polyacetylene [341, 342]. These data suggest that the degree of cross-linking must be limited to less than 1%. There is evidence for quite high levels of sp^3 carbons in polyacetylene produced on the Luttinger catalyst (10–12%) which contains $NaBH_4$ and could hydrogenate the chain [343]. Other workers have reported up to 4 mol% $-CH_2-$ units in polyacetylene, but this may be associated with high temperatures of isomerization from cis to trans [344].

Gibson [345] claims that about 5% of cis-units can be detected in polyacetylene by ir even after extended periods of transformation to trans-polymer, although all-trans Shirakawa polyacetylene can be synthesised if the reaction is performed at a high enough temperature [346]. Chien and Schen [324] also suggested that isomerization of high molecular weight polyacetylenes can trap residual cis-units and claimed that these can react to give sp^3 units on prolonged heating. We have given detailed attention to the possibility of chain defects in polyacetylene produced by the Durham presursor route and find no evidence for sp^3 carbons, uneliminated side groups or residual cis-linkages [347, 229].

The synthesis of poly-p-phenylene by oxidative coupling of benzene might be expected to produce at least some ortho-linkages. Brown et al. [334, 348] have looked at this polymer by Laser Desorption Fourier Transform Mass Spectroscopy. They suggest that samples formed by metal halide — oxidant systems are complex mixtures of oligomers including some polynuclear structures. Polymerization with metal chlorides leads to termination of the chain growth by chlorination of the end rings and reduces the formation of polynuclear centres. Duroux et al. [349] suggested that the preparation-dependent dielectric conductivity of poly-p-phenylene arises from a halogen dipole.

Kovacic polyphenylene is brown with about 1 spin/chain detectable by ESR. Yamamoto polymer is yellow with a shorter chain length and fewer spins. In Kovacic polymer the spins and colour may both be due to polynuclear species. Polyphenylene produced from the poly(dihydrocatechol) precursor [249] is also yellow, but has a high molecular weight, of the order of 10^5. It contains about 15% o-linkages, and the aromatization procedure may leave a high level of twists in the chain originating from the flexible precursor. This material dopes only to low levels of conductivity with sodium naphthalide (6×10^{-3} S cm^{-1}) and iron chloride (1.5×10^{-2} S cm^{-1}) but reaches a level comparable to Kovacic and Yamamoto polyphenylenes with AsF$_5$ (10^2 S cm^{-1}).

There is evidence from many sources to suggest that polypyrrole is predominantly linked in the 2,5-positions, including the fact that the dominant product when 2-methyl pyrrole is electrolysed is the dimer and that only low molecular weight products are formed from 2,5-disubstituted pyrroles [350]. Nazzal and Street [337] showed that the molecular weight of poly(3,4-dimethylpyrrole) is of the order of 100,000, by measuring the tritium content of polymer formed by the reaction of the monomer tritiated in the 2- and 5-positions. When they applied this approach to pyrrole itself, they found a molecular weight of around 500. On the basis that this is too low for the other properties of the polymer, they concluded that there is residual tritium due to structural disorder in the polymer chain, possibly β-links or saturated carbon atoms. Osterholm et al. [351] used ^{13}C CPMAS nmr spectroscopy to demonstrate that polythiophene and poly(3-methyl thiophene) are both predominantly 2,5-linked. However, it should be remembered that a few percent of cross-linked units could suffice to modify the crystallinity, mechanical properties and solubility of these polymers, and such low levels of cross-linking could be hard to detect chemically or spectroscopically. Roncali and co-workers [140] have shown that the conductivity of poly(3-methylthiophene) decreases with increasing monomer concentration, which they attribute to higher levels of oligomeric species coupling to growing polymer chains. The reactivities of the 4- and 5-positions are more similar in the oligomers, and so 4,5-coupling, and loss of conjugation, is more likely. This was confirmed by the shift of the optical absorption to lower wavelengths in polymers made from terthiophene.

Brown et al. [352] have recently emphasised the role of defect structures in heterocyclic polymers. They point out that the reported doped conductivities of these polymers may vary by as much as six orders of magnitude depending on the preparation procedure. They have applied the laser desorption method, discussed earlier for polyphenylene, to a range of polyheterocycles. Unlike polyphenylene, there was evidence for incomplete desorption and rearrangement of evaporated molecules. The results show that polymers prepared by Grignard coupling vary in their extent of bromination, the nature of the terminal species and the extent of formation of cyclic, polynuclear contaminants.

4 Polymer Morphology and Mechanical Properties

The fundamental difference between polymers and other materials is the high molecular weight of chain polymers. If we consider the alkane series, there is a progressive

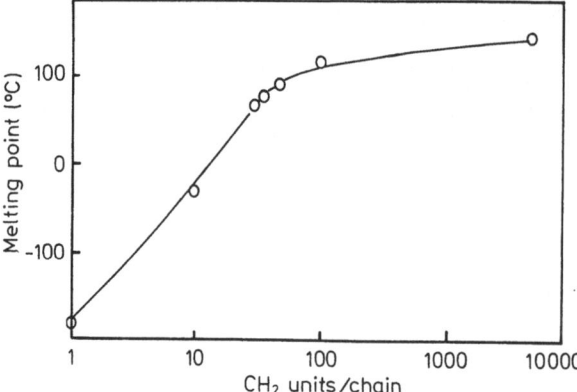

Fig. 10. Effect of chain length on the melting point of linear polyethylene

change in all properties with increasing molecular weight. Thus, as shown in Fig. 10 the melting point increases with chain length and starts to level off at around 100 carbons. However a chain length of 100 carbons is not sufficient to give the material the properties of toughness, elasticity and high melt-viscosity that we normally associate with polyethylene.

As is shown in Fig. 9 the structure and mechanical behaviour of the material do not really become typical of polymers until chain lengths of around 1000 carbon atoms and these properties continue to change up to the highest attainable chain lengths of around 10^6 units. At the risk of oversimplifying we can say that the plastic properties of polymers arise from the presence of chains which are flexible and sufficiently long to become entangled. In addition, intramolecular forces are usually very much larger than intermolecular forces. Thus, if we are to interpret the idea of "conducting polymers" as materials which are both conducting and plastic, the assessment of chain length and chain stiffness become very important.

To a synthetic chemist the concept of a polymer is rather different. The term is often used to describe molecules formed from quite small numbers of monomers units but which resist crystallization and purification owing to the range of chain lengths present. Such materials will normally form brittle solids which are either amorphous or have a low level of crystalline order.

Many polymers are not simply linear, or even branched, chains but are cross-linked into a three-dimensional network. If the cross-linking is light, with many chain units between the cross-link points, the characteristic polymer properties of toughness and flexibility are maintained although viscous flow in the liquid state becomes impossible; rubbers are such cross-linked liquids. As the cross-link density increases, the mobility of the chain sections between cross-links is lost and the material becomes a brittle glass. Thus the inorganic glasses can be considered as very heavily cross-linked polymers with very little plasticity while the cross-linked resins, such as epoxies, form an intermediate state with the linear polymers at the other extreme. Against this background, we need to consider whether the conducting polymers can be considered as true polymers or should really be described as conducting organic glasses.

A process of simultaneous polymerization and precipitation occurs for any polymerization in which the polymer is insoluble in the reaction medium, and is thus a feature of almost all syntheses of conducting polymers. Where the polymer is amorphous, one expects to find globular powdery morphologies but more interesting morphologies are often seen when the polymer is crystalline. Fibrillar structures are familiar in polymerizations of ethylene and propylene on both soluble and heterogeneous catalysts [353]. It is unclear whether the chains first form, and then precipitate from solution, or are formed directly on the fibril surface. Fibrillar structures are often associated with polymerization at higher temperatures or in better solvents where partial solubility of the polymer is expected and chains can interact in solution. Powders tend to be formed under conditions where no solubilization would occur during chain growth. Low catalyst concentration (so that chains grow in isolation), heterogeneous catalysts, polymerization in poor solvents or formation of short chains also tend to give a powder rather than fibrils [354].

4.1 Polyacetylene

As prepared by the method of Shirakawa et al. [2, 3], polyacetylene is a free-standing film. On closer examination, its density is found to be around $0.4\,g\,cm^{-3}$, only about 30% of the value ($1.16\,g\,cm^{-3}$) predicted from X-ray analysis, and electron microscopy reveals complex morphologies.

The materials used in most current research are irregular mats of highly crystalline fibrils with diameters of around 10 nm, so that the films are characterised by a very high surface area (around $60\,m^2\,g^{-1}$), a problem in some potential applications and an asset in others. The morphology of polyacetylene is sensitive to the conditions of preparation and to ageing and was the subject of much heated discussion in the early development of polyacetylene.

The effect of polymerization conditions on polyacetylene morphology has recently been examined by Abadie et al. [355], and Chien [6] has discussed the morphology in great detail. As prepared on a reactor wall the film has a dense, shiny side facing the wall and a more open, dull side facing out. The shiny side of the film has a higher concentration of catalyst residues which are not removed by washing [356]. A change of solvent from toluene to mixtures with THF gave a more globular structure as the THF content increased, possibly reflecting a less extended polymer conformation in THF; globular films were also obtained from anisole [357]. In association with the fibrils in many polyacetylene preparations is seen a particulate morphology which can be partly removed by washing in methanolic HCl. These particles may be small lamellar crystallites which form simultaneously with the fibrils [358]. From washing experiments, it seems that the film comprises a fibrillar network with the chain axis along the fibrils and adsorbed lamellar particles where the chain axis is in the lamellar thickness direction. The ratio of the two forms depends on the polymerization conditions. Films produced by polymerization in silicone oil [15, 16] have higher density ($0.6\,g\,cm^{-3}$) but retain the fibrillar morphology.

During a normal polymerization of acetylene, in addition to the film which forms on the reactor wall, there is also a gel which forms on the solution surface, plus powder which forms in the bulk solution. According to Karasz et al. [359], the globular morphologies sometimes seen, arise from partial melting if the high heat of polymerization is not dissipated; although polyacetylene has no melting point, small units can apparently be melted or solubilized during polymerization.

Chien et al. [323, 324] have followed morphology changes with molecular weight, controlled by ageing of the catalyst. They found that fibril diameter decreased from 70 nm to 10 nm as the molecular weight decreased from 870,000 to 500. The low molecular weight polymer forms brittle films or powders. Chien et al. [360] have also shown that macroscopic ribbons of polyacetylene can be grown by polymerization in a stirred solution, in a manner similar to the formation of polyethylene fibres from stirred solutions, described by Pennings and his co-workers [361].

Polyacetylene formed on other catalysts is generally more powdery and the films are weaker. The polymer formed on the Luttinger catalyst is of lower crystallinity (60%) and has a higher sp^3 content of 10% versus <2% for Shirakawa polyacetylene [362]; the final conductivity of iodine-doped material is also lower. When first discussed by Lieser et al. [24], these catalysts seemed to give morphologies rather different to those formed by the Shirakawa system. Chien et al. [363] suggested that the apparent globular morphology is an artefact caused by the difficulty of removing high concentrations of catalyst residues from the polymer and have shown that prolonged treatment of Luttinger polyacetylene with methanolic HCl leads to a fibrillar morphology, similar to that of the Shirakawa polymer.

While polyacetylene is quite tough in the *cis*-form, films which have been isomerised to *trans* are generally brittle [364]. One study has reported mechanical measurements on a range of substituted polyacetylenes [365]. These polymers resemble the vinyl polymers in that long alkyl side chains render them soft and tough while poly(phenylacetylene) is hard and brittle. The glass transition temperatures are all about 200 °C as this seems to be determined more by the backbone stiffness than by the side groups. For *cis*-polyacetylene the Young's modulus is 200 MPa and the elongation to break more than 200% but these properties may reflect the open fibrous structure rather than intrinsic properties.

Ultrasonic disintegration of Shirakawa films leads to a suspension of fibrils with a hydrodynamic radius of 300 nm [14]. Drying of the suspension gives a film resembling the original but weaker. This certainly suggests that the fibril is the basic structural unit in Shirakawa polyacetylene.

Many attempts have been made to change the morphology of polyacetylene and to induce orientation, either by mechanical treatment of the polymer or by appropriate modifications to the polymerization reaction. Chien et al. [360] described oriented polyacetylene ribbons produced in the shear field near the tips of a rotating stirrer bar onto which the Shirakawa catalyst was absorbed. Meyer [366] polymerized acetylene in the shear field of a Couette rotating cylinder viscometer and was able to obtain polyacetylene ribbons with some degree of orientation, at the expense of a reduction in overall crystallinity. Aldissi [367] polymerized acetylene with a Shirakawa catalyst dissolved in a nematic liquid crystal solvent, which was aligned in a magnetic field during the polymerization. X-ray diffraction showed some orientation, and conductivity studies on doped samples showed a four-fold higher conductivity parallel to the chain axis and an 85% reduction in the activation energy for conduction. Akagi et al. [368] recently showed that films of Shirakawa polyacetylene prepared under these conditions are oriented and can be doped to very high conductivity.

The same authors [369, 370] also obtained similar results if the liquid crystal solvent was aligned by flow during the polymerization. They showed that the polymerization conditions lead to alignment of the fibrils within the polymer mass and of the chains within the fibrils; polymers produced in this way could also be doped to a conductivity of 10^4 S cm^{-1} [371]. The morphology of polyacetylene produced by polymerization in a liquid crystal solvent, aligned both magnetically and by flow, has been studied by Montaner et al. [371]. They show that the polymer film is made up of very long fibrils built from microfibrils. In one fibril, the orientation of microcrystalline domains with respect to the fibril axis is very well defined, whilst the orientation of the different fibrils in the sample spreads over 20°.

Woerner et al. [373] produced polyacetylene with locally oriented regions and an optical anisotropy of $2 \times$ by polymerization on crystals of biphenyl. Yamashita and co-workers [374, 375] have recently reported epitaxial polymerization of acetylene on crystals of anthracene, naphthalene and biphenyl where fibrils of *cis*- or *trans*-polymer formed, crystallographically aligned with the substrate. Fincher et al. [376] produced a $3 \times$ extension which gave a $4 \times$ optical anisotropy.

Several attempts to induce orientation by mechanical treatment have been reviewed [6]. *Trans*-polyacetylene is not easily drawn but the *cis*-rich material can be drawn to a draw ratio of above 3, with an increase in density to about 70% of the close-packed value. More recently Lugli et al. [377] reported a version of Shirakawa polyacetylene which can be drawn to a draw ratio of up to 8. The initial polymer is a *cis*-rich material produced on a Ti-based catalyst of undisclosed composition and having an initial density of 0.9 g cm^{-3}. On stretching, the density rises to 1.1 g cm^{-3} and optical and ir measurements show very high levels of dichroism. The (110) X-ray diffraction peak showed an azimuthal width of 11°. The unoriented material yields at 50 MPa while the oriented film breaks at a stress of 150 MPa. The oriented material, when iodine-doped, was 10 times as conductive (2000 S cm^{-1}) as the unstretched film. By drawing polyacetylene as polymerized from solution in silicone oil, Basescu et al. [15, 16] were able to induce very high levels of orientation and a room temperature conductivity, after doping with iodine, of up to 1.5×10^5 S cm^{-1}.

A particularly interesting property of Durham polyacetylene is that it can be stretched to draw ratios of up to 20 during the transformation, to yield a polyacetylene sample with high levels of orientation. This effect was reported by Bott et al. [378] for thin films in the electron microscope and then by Leising et al. [379], who drew single fibres of polyacetylene to a highly oriented *trans*-state with a density of 1.06 g cm^{-3}.

The wetting properties of polyacetylene have been studied by Schonhorn et al. [380] who measured a critical surface tension of 51 mN m^{-1}, considerably higher than for other hydrocarbon polymers. This value was attributed to oxidation of the surface as no change was observed on further oxidation. Treatment of polyacetylene with aqueous potassium permanganate renders it hydrophilic, reduces the contact angle for water from 72° to 10° and renders the structure more water-permeable [381]

On doping of polyacetylene by iodine, the fibrils swell but the original fibrillar structure remains intact and no segregation of iodine on the surfaces is seen [382]. On swelling with metal halides, densification appears to occur preferentially near the film surface, which was interpeted as a sign of staging [383]. Dopant profiles measured through films doped with WCl$_6$ and MoCl$_5$ showed that doping only became uniform through the film at 10 mol % dopant [384]. Similarly, ESR studies of lithium doping from solutions of the Li-benzophenone complex in THF showed both sharp and broad peaks after short periods of doping, suggesting that doping is inhomogeneous [385].

4.2 Poly(*p*-Phenylene)

As formed by the Kovacic route, polyphenylene is a fine brown powder made of rod-like particles of about 0.2 μm diameter and with a surface area of 50 m^2 g^{-1}, which would suggest a particle radius in the region of 40 nm [386]. This structure must arise from precipitation during the polymerization process, but it is not clear whether polymerization continues on the particle surface. The absence of high molecular weight species suggests not. The morphology is then determined by the process of separation of the polymer from solution in benzene. The particle size and crystallinity of material formed on FeCl$_3$, is lower than that formed on molybdenum chlorides [387]. Sexiphenyl, a lower chain length model, can form liquid crystalline phases [388]. After ultrasonic disintegration, Kovacic polymer appears as entangled fibres in transmission electron microscopy [389]. The fibres are 40 nm diameter and contain small crystallites of about 4 nm with the chain axis parallel to the fibre axis. On doping the fibres increase in size to 60 nm. A scanning electron microscopy study also shows rods of about 500 μm long by 100 μm diameter which are composed of fibrils of about 0.1 μm diameter [390]. As pressed into a pellet, polyphenylene has a compressive strength of 14.2 MPa, a modulus of 150 MPa and fractures at a strain of about 10 % [327]. The low value for modulus and high elongation suggest that the samples were poorly packed powders. A tensile strength for pressed and sintered powder has been reported as 34 MPa, with a modulus of 3.5 GPa and an elongation to break of 1 % [39].

Polyphenylene formed from simultaneously doped and polymerized crystals of terphenyl has the form of a macroscopic slab. The final conductivity with AsF$_5$ doping is 60 S cm^{-1} along the chains and 40 S cm^{-1} perpendicular to the chains, less of a difference than would be expected on a simple chain conduction model [61].

4.3 Electrochemically-Synthesized Polymers

Polypyrroles formed from anhydrous acetonitrile solutions with tetraethylfluoroborate counter-ions are rough and dendritic on the scale of 5 μm. With 1 % water in the solution, the film becomes smooth to a scale of about 100 nm [392]. This sensitivity to small amounts of impurity is rather reminiscent of electroplating processes. The density of the films is 1.48 g cm^{-3}. Films can also be produced by chemical reaction at solid surfaces or at liquid-liquid boundaries but have poor mechanical properties, presumably due to low molecular weight [393]. Electrochemically-synthesised films have a strength of 44 MPa and an elongation to break of 5 %. By preparation of films at −20 °C, an elongation to break of over 100 % could be obtained which gave a conductivity parallel to the draw direction (1000 S cm^{-1}) which was three times that of the starting film [394]. Niwa and Tamamura report increasing the elongation to break to 22 %, but a reduced fracture strength of 12 MPa in PVC-polypyrrole

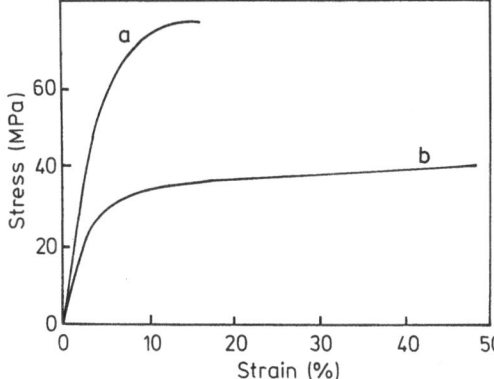

Fig. 11. Stress-strain behavior of a) polypyrrole doped with *p*-toluene sulphonate and b) the same polymer plasticised with acetonitrile. Reproduced with permission from Wynne KJ, Street GB (1985) Macromolecules 18: 2361

composites whose conductivity was in the same range as for polypyrrole (5–50 S cm^{-1}) [295]. A composite of polypyrrole with a vinylidenefluoride copolymer showed an elongation to break of 120% [297].

A study of the dependence of mechanical properties of polypyrrole-*p*-toluenesulphonate on conditions of preparation showed that films prepared at low current density were stronger with a fracture strength of up to 65 MPa and a Young's modulus of 4 GPa [395]. Naarman has suggested that flexibility and smoothness are both a function of the cation. He describes films with a crystallinity of 15% and a surface area of 5–50 m^2 g^{-1}. This area is much higher than the geometric area of the films and corresponds to a spherical particle diameter of 0.8–0.08 μm [396]. Buckley et al. [397] studied the influence of preparation conditions and counter-ion on the mechanical and electrical properties of polypyrrole films. They showed that the degree of ordering is low and determined by the anion and the synthesis conditions and that increased ordering gives higher conductivities. The fracture strength of the films decreased with increased current density during synthesis, the highest values being about 65 MPa. At the same time, increased current density gave films with lower flotation density. Fig. 11 shows the mechanical properties of a *p*-toluenesulphonate doped polypyrrole.

Glatzhofer et al. [128] describe the reduction of conductivity of polypyrrole films, doped with poly(styrene-*p*-sulfonate), as increasing concentrations of dioxane are added to the aqueous electrolyte. They claim that the solvent induces a conformational change in the counterion, which modifies the film morphology.

Tourillon [136, 398] has reported that polythiophene films formed electrochemically are smooth up to 200 nm thick, after which they become rough and eventually powdery, at thicknesses above a few μm. The films have a density of about 1.5 g cm^{-3}. Undoped films of polymethylthiophene show fibrils of 25 nm diameter which swell to 80 nm on doping to 25 mol% with perchlorate conter-ions.

The electrical conductivity of polythiophene is much higher in the plane of the film (about 10 S cm^{-1}) than perpendicular to the film (10^{-4} S cm^{-1}). X-ray diffraction shows a slight preferential chain orientation in the plane of the film and the structure appears to be layered under some preparation conditions [399].

Satoh et al. [400] have reported yield strengths of 80 MPa for undoped polythiophene, increasing to 170 MPa after drawing to 60%. The redoped drawn film had a conductivity of 150 S cm^{-1} along the draw direction, 2.5 times the conductivity of the undrawn films. The drawability increased with the electrolyte concentration in the polymerization solution. Ito et al. [401] found fused-globular morphologies in polythiophenes and showed that the tensile modulus and strength of the films decrease with increase in the current density employed in the polymerization; at a given current density the mechanical properties of the undoped film were better than those of the doped film.

Poly(p-phenylene), synthesised electrochemically, is generally reported as a 'smooth' film, although published micrographs indicate a fused globular morphology [157–167].

As with other polymers, the morphology of electrochemically-synthesised poly-aniline is sensitive to the counter-ion and the preparation conditions. Wang et al. [402, 403] showed that the morphology is dependent upon the form of the initial film laid down on the bare platinum electrode surface, which may be an open, fibrillar structure or a smoother, more condensed film, depending on the acid used to provide the counter-ions. Similarly, Chen and Lee [404] claimed that use of HBF$_4$ gives a fibrillar film, whereas HCl or H$_2$SO$_4$ give smoother but granular films. Kitani et al. [405] have reported that synthesis of polyaniline in aqueous solutions of either HBF$_4$ or HClO$_4$, followed by cathodic undoping, can give thick (>0.5 mm), spongy films which can be removed from the electrode and are tough, with an elongation at break of 40%. This toughness is at the expense of conductivity, because the polymer returns to its brittle state on redoping.

4.4 Discussion

The intractable properties of conducting polymers do not allow much manipulation of the morphology, so that most studies are restricted to effects of different nascent (as-polymerized) morphologies on properties. The development of morphology during polymerization is not well understood; Wunderlich [406] has reviewed the underlying principles. All crystalline polymer morphologies are non-equilibrium structures in that there are no large single crystals of polymers except in special cases of solid state polymerization, such as the diacetylenes. Rigid, insoluble chains like conducting polymers have even less chance to rearrange during formation and so structure must reflect the polymerization dynamics.

The development of morphology during electrochemical polymerization is also black art, although it is of importance in determining whether films or powder results. Presumably, principles must apply which are common with those governing electroplating, but there has been no detailed study of this area. The initial deposition of polypyrrole or polythiophene gives a thin uniform film, which then becomes unstable so that the growth front breaks up into a more powdery morphology. The source of orientation in polythiophene films is unclear. The electric fields are not large enough to cause molecular alignment, but the phenomenon may be due to shrinking during drying.

The mechanical properties of conducting polymers can be compared with those of other crystalline and glassy polymers as summarised in Table 2. For polyacetylene and polypyrrole, the strength and elongation to break are quite respectable, suggesting that the material is a high polymer and not a low molecular weight or heavily cross-linked resin. In contrast, the properties of polyphenylene are significantly worse. The mechanical properties of all conducting polymers, would be expected to depend

Table 2. Comparative mechanical properties of polymers

Polymer	Tensile modulus GPa	Tensile strength MPa	Elongation to break %
Polyethylene	1.2	27	450
Polystyrene	3.3	50	3
Phenolic resin	7	55	1
Nylon 6	1.3	61	250
Nylon 6 drawn fibre	4.4	690	25
Ultradrawn polyethylene	100	3000	10
Polyacetylene		50–150	
Polyphenylene	0.15	14	10
Polyphenylene	3.5	34	1
Polypyrrole	44	5	
Polypyrrole	4	65	
Polythiophene	80–170	> 60	

on the dopant content and on plasticization by residual solvent, but there seem to have been no systematic studies of these effects.

One interesting aspect of the precursor polymers is the ease with which they can be drawn to highly oriented conducting film or fibre. Recent research on ultra-drawing of polyethylene has shown that very high orientations can be obtained on drawing at close to the melting point [407] or on disentangled gels [408]. Apparently the transformation of the precursor during drawing allows a similar level of controlled slip.

5 Electronic Structure and Relation to Properties

The relationship between chemical structure and electronic properties in conjugated-chain polymers has been extensively reviewed [409]. In this section we only briefly discuss the issues which seem relevant from the point of view of polymer science.

The availability of electronic states, arising from the conjugated π-electron system and capable of facile oxidation and reduction, is the key to the doping behaviour and electrochemistry of conducting polymers but is also responsible for much of their environmental instability. For any conjugated structure there are two critical properties which it is possible to calculate, from a consideration of the configuration of the π-orbital system, the band gap and the ionization potential. Both *ab-initio* and other calculations have given values in reasonable agreement with experiment [410]. The ionization potential is important for the ability of oxidizing agents (dopants) to remove electrons from the π-orbitals and to the stability of the polymer to oxidation in air. The band gap will determine the number of electronic defects in the undoped polymer and its optical spectrum. Some of these defects will be neutral, such as solitons (unpaired spins) in polyacetylene, other defects will be

PREDICTED SMALL BAND GAP POLYMERS

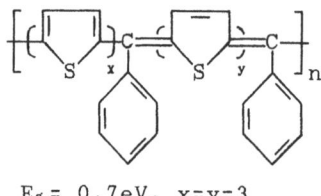

Polyisothianaphthene
$E_g = 0.54$ eV

Polythieno[3,4-c]thiophene
$E_g = 1.02$ eV

Polyisonaphthothiophene
$E_g = 0.01$ eV

OBSERVED SMALL BAND GAP POLYMERS

$E_g = 0.7$ eV, x=y=3

Fig. 12. Proposed and known polymers with small band gap

charged (polarons) and so can act as current carriers. The conductivity of the undoped polymer will depend on the number of charged defects and their mobility.

In principle, it should be possible to design and synthesise a polymer with a sufficiently small band gap that it would be conducting without doping. Bredas[411] has discussed theoretical calculations of the band-gap of polymers based on polyisothianaphthene. A number of polymers have band-gaps predicted to be less than 1 eV, with polyisonaphthothiophene having a predicted band-gap of almost zero. Jenekhe[412] has made polymers with a band-gap of 0.73 eV, based on polythiophene with alternating aromatic and quinonoid units (Fig. 12).

The mobility of charge carriers can be measured by photoconductivity measurements, or by studies on diodes, which also allow the number of carriers to be estimated. Values for carrier mobility in polyacetylene (holes in oxidised polymer) are $< 10^{-4}$ m^2 V^{-1} s^{-1}, which compares with 0.05 m^2 V^{-1} s^{-1} for holes in silicon. This very low mobility leads to a low conductivity. Doping serves to add or remove electrons to or from the π-electron system of the polymer. This creates large numbers of charge carriers and the conductivity rises, although there may be some drop in the mobility. At about 6 mol% dopant most of these polymers become metallic; the conductivity no longer shows the large activation energy associated with hopping but decreases with increasing temperature, and the esr signal shows the characteristics of a metal. The implication is that, at this level of dopant, the wavefunctions associated with the charged states start to overlap. Nonetheless, the conductivity in the metallic state is still dependent on the polymer and the dopant, over a range from 1–10^5 S cm^{-1}.

Theoretical consideration of a perfect crystalline conjugated chain is a viable basis for comparing polymer structures and trying to understand their doping and conductivity, although it does not necessarily bear much relation to reality. In the early stages of the development of conducting polymers it was widely believed that conduction could only arise if the polymer chains are highly conjugated, very long and regularly packed into close, parallel arrays in the doped state. With time, it has been realised that most of these requirements are much less stringent than was at first imagined. Approaches based on perfect crystal models are limited when applied to real polymers, which are highly defective solids, frequently amorphous and rarely very pure. One consequence of the defective nature of the real polymers is that the band-gap is not well defined. The introduction of defects and impurities puts electronic states into the band-gap with ill-defined energies. Thus the ionization potential is both lower than the theoretical value for the perfect chain and also changes continously with the degree of ionization. The same is true of the related quantity, the electro-chemical oxidation or reduction potential. Although it is possible to discuss the doping behaviour of conducting polymers in terms of their electrochemical potentials [413], these are again ill-defined. A conventional oxidation potential cannot be defined for a conducting polymer, since it is impossible for the oxidised and reduced states to coexist in the same solid phase at unit activity. Additionally, the oxidation state of a polymer changes continuously with extent of oxidation, not in integral steps. Comparison of the energetics of electrochemical and gas-phase doping is also complicated by the energetics of interaction of the polymer with its counter-ion; in the liquid phase the counter-ion may be nominally the same as a gas-phase sample but is likely to be solvated.

We have seen above that there is reasonable evidence to suggest that polyacetylene chains, whether formed by the direct polymerization of acetylene or via precursors, can have molecular weights corresponding to many hundreds of conjugated double bonds, with little evidence that these sequences are interrupted by sp^3 defects. The length of the molecular chain is apparently largely irrelevant to conductivity, since Schen et al. [322] showed that the conductivities of both undoped and iodine-doped samples of Shirakawa polyacetylene are independent of molecular weight in the range 400 to 870,000. This suggests that transport of charge carriers between chain ends is not an important component of the conductivity or, more precisely, that conduction along the chains is not a major factor in electron transport. Photo-induced conductivity studies on oriented polymer [414], show that most conduction occurs by hopping between chains, not by transport along chains.

In principle, a perfect polyacetylene chain would have a completely conjugated sequence of *trans*-double bonds along its whole length and we would thus expect conjugation lengths to be equal to the molecular length. Conjugation lengths in poly-enes have typically been estimated by uv spectroscopy or, with greater resolution and sensitivity, by resonance Raman spectroscopy. This latter method has been applied both to degraded PVC [415] and to carotenoids [416]. The interpretation of Raman intensities to yield conjugation length distributions is controversial. However, there is reasonable agreement that the spectrum of Shirakawa polyacetylene corresponds to a distribution of short conjugation lengths plus sections having 50 to 100 double bonds [417], so that the conjugation length appears to be significantly shorter than the chain length. In marked contrast, Durham polyacetylene has a particularly short conjugation length, when fully converted to the *trans*-isomer, the Raman dispersion corresponding to sequences of 20 to 30 double bonds [418]. The esr behaviour of Durham polymer also supports this view in that the spins (solitons) are considerably less mobile at room temperature than they are in Shirakawa polyacetylene [419]. The exact nature of the conjugation-limiting defects in polyacetylene is uncertain. In the case of the Durham material, we believe that it is related to the constraints on mobility limiting chain straightening during conversion from the randomly coiled precursor polymer. The result is polyacetylene still approximating to the original random coil, but with straight sections connected by kinks, so that the defect is conformational, retains the bond-alternation symmetry of the chain and may be 'transparent' to the charge carriers [420]. Others [421] prefer to view the polymer as cross-linked and analogous to degraded PVC, despite the fact that the polyacetylene can be doped to metallic conduction whilst PVC cannot.

Although there seems little argument that the conjugation length of a polyacetylene chain, as measured spectroscopically, is shorter than the chain length of the polymer and that the conductivity of a polyene chain is independent of its chain length, there is much less agreement on whether the conductivity is dependent upon the conjugation length. Thus most preparations of Shirakawa polyacetylene can be doped to conductivities within a factor of two of 1000 S cm^{-1}. Durham polyacetylene, as normally prepared, is amorphous, has a much shorter conjugation length and its conductivity is about is about one order of magnitude lower than for Shirakawa polyacetylene, both before and after doping. This is in rough agreement with other estimates which suggest a dependence of conductivity on conjugation length to a 3.2 to 3.5 power [422]. Various annealing treatments do not have much effect on conductivity or conjugation length in Durham polyacetylene, though orientation does increase the conductivity of the undoped polymer along the drawing axis [423].

There have been a number of studies aimed at modifying the conduction properties of polyacetylene by incorporating chain defects. Chien and Babu [424, 425] have described copolymers of acetylene with carbon monoxide and have shown that they can be doped to high conductivity. Since the carbonyl groups should interrupt the conjugation, they conclude that conductivity requires only short conjugated sequences. They also point out that the spin content of these copolymers is very much lower than that of Shirakawa polyacetylene, which has roughly one spin, identified with the soliton, per chain. The spin mobility is not reduced in the copolymers. On this basis they question the relevance of solitons to conductivity in polyacetylene. However, the postulated structures of these copolymers and the proposed mechanisms for doping have proved very controversial [426, 427].

Schafer-Siebert et al. [428] deliberately introduced sp^3 defects into Shirakawa polyacetylene by *p*-type doping, followed by hydrolysis. They found that the introduction of 15 % of such defects leads to a drop of three orders of magnitude in the conductivity achievable with iodine doping. Similarly, as discussed in Sect. 9.1 below, the inclusion of sp^3 defects into the polyene chain by oxidative degradation reactions also dramatically reduces conductivity of both doped and undoped samples. Other workers have sodium-doped polyacetylene and then converted the Na sites into (CHD) groups by reaction with deuterated alcohol to leave up to 17 % CHD groups in the chain. The conjugation length, measured by optical absorption, drops more slowly than would be expected for a uniform distribution of sp^3 carbons [429]. The spin content changes little on treatment but, with 17 mol% (CHD), the conductivity after iodine doping is 1/4000 of that of the untreated material. In all these systems of modified polyacetylene, there is no certainty that the defects are uniformly distributed along the chain. Copolymerization may lead to "blocky" structures, while post-treatment may give localised reaction.

Baughman and Shacklette [423] have proposed a theoretical model for conduction in which the carriers are free to move along the chain between defects ("blocks"), but the conductivity will be limited by the rate of hopping between chains, or of intra-chain hopping around blocks. Based on the anisotropy of conductivity in drawn polymers and its temperature dependence, they argue that conductivity is dominated by interchain hopping and derive an l^3 dependence of conductivity on conjugation length, in agreement with experiment.

In summary, the conductivity of polyacetylene is correlated with conjugation length, but conjugation length does not vary simply with chemical modification of the chain. Spin density in *trans*-polyacetylene has little to do with conductivity in the undoped or doped states. These observations can be rationalized on the basis that sufficient conjugation is required to stabilize the charge-carrying states, but charge transport is controlled mainly by inter-chain hopping. Since most of the forms of polyacetylene have similar chain packing, there cannot be large variations in the extent of inter-chain interactions.

The recent observations of very high conductivities in drawn, highly oriented, polyacetylene produced in silicone oils [15, 16] and in liquid crystal solvents [372] suggest that very high intra-chain conductivity can be attained if high levels of perfection are reached. Naarmann and co-workers attribute the high conductivity of their polyacetylene to a zero level of sp^3 defects, and to a higher density. According to the Baughman and Shacklette model, the higher conductivity would arise from a greater conjugation length, allowing more transport of charge along chains. It is well known that conductivity is very dependent on the polymerization conditions [430] and it may be that the balance of polymerization kinetics and morphological development can have a profound effect on chain folding and defect levels, as is found in other polymers (see Sct. 4.4 and 7.7).

Almost all conductivity studies to date have apparently been dominated by the low inter-chain hopping mobilities. Hopping around defects within the chain is apparently not important as there is no molecular weight dependence of conductivity. The conjugation lengths are thus normally long enough to stabilise the charge carriers but too short to allow extensive intra-chain mobility. When defects along the chain are removed, the intra-chain mobility in polyacetylene rises dramatically. The implication is that much higher intra-chain mobilities are also attainable for other conducting polymers. Morphological factors play a large role in determining the mechanical properties of polymers and recently much study has gone into the production of very high modulus fibres by drawing to high ($>20\times$) extensions such that the backbone stiffness dominates the fibre stiffness. Modulus along the backbone can be two orders of magnitude higher than the modulus perpendicular to the chains. It is now becoming apparent that conductivity shows similar anisotropy.

6 Thermal Response

One focus of the interest in polymer structure-property relationships has been the prediction of glass transition points and melting points from chemical structure [431]. In general, it is expected that both T_g and T_m will respond in similar ways to structure with T_g being typically of the order of $1/2$–$1/3$ of T_m in absolute temperature. Influences on T_m can be related to changes in chain stiffness and to changes in intermolecular forces. An increase in chain stiffness can be seen as a decrease in the number of low energy conformations available to an isolated chain and thus a decrease in the entropy change on melting when the chain goes from a single repeated conformation to a random coil. Hence, the very stiff chain of polytetrafluoroethylene where the large fluorine atoms interfere in gauche conformations of the chains, has a melting point of 330 °C while polyethylene is relatively more flexible and has a melting point of 140 °C. Strong intermolecular forces, such as hydrogen bonding in the nylons, further lower the energy of the crystalline state with respect to disordered states and so increase the melting point. Hence, the melting point of nylon 66 is 266 °C, 125 °C higher than polyethylene.

Similar arguments can be developed for the glass transition, with the difference that it is the rate of transition between states that is more important than the energies of the various minima and so it must be the height of the energy barriers to rotation which determines the glass transition temperature.

The chief characteristic of the conducting polymers is conjugation along the chains and hence great chain stiffness. This would be expected to make both T_g and T_m very high. In addition, the extensive delocalization of charge along the chains may lead to strong intermolecular forces. The case of polyacetylene is quite clear, any rotation out of the trans-state requires the disruption of the π-bonding system and an energy increase of the order of 100 kJ mol^{-1}. Further, the *cis*-conformation has an energy 10 kJ mol^{-1} greater than the *trans*-state so that the heat of fusion would be at least this large, compared to 4.0 kJ (mol–CH$_2$–)$^{-1}$ in polyethylene. The glass transition would be similarly very high. In fact, there is no melting known for polyacetylene. It is conceivable that the onset of the *cis-trans*-change in amorphous (Durham) *cis*-polyacetylene could be regarded as the onset of chain mobility and hence related

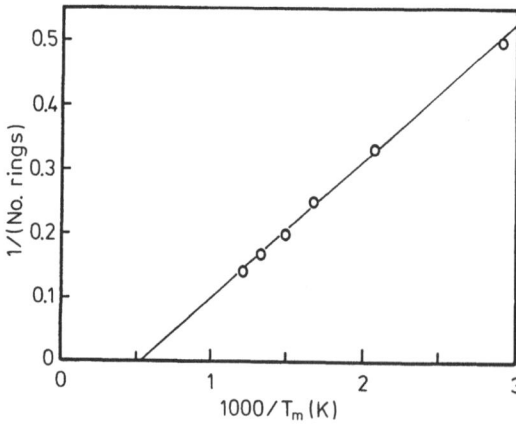

Fig. 13. Dependence of melting point on chain length for polyphenylenes

to the glass transition. In part, this interpretation depends on whether the conformational change is viewed as a zipping process, involving a propagating defect [432], or as a random process [433].

Polyphenylene melting points have been recorded by Speight et al. [47], for short chains. Biphenyl melts at 70 °C while sesquiphenyl melts at 545 °C; extrapolation to 20 units would give a melting point of around 1580 °C (see Fig. 13). Until recently, the only conducting polymer for which a melting point has been recorded is polyphenylene sulphide (295 °C) but present evidence suggests that this polymer converts to a ladder form [434] on doping and thus does not itself belong to the group of conjugated polymers but to the group of precursor polymers.

Against this background of infusible conducting polymers, the development of the soluble polythiophenes is most interesting. Glass transition temperatures have been reported as 48 °C for poly(3-butylthiophene) and 145 °C for poly(3-methylthiophene) [261]. These polymers also show crystallinity in films and can be crystallized from solution. Solution studies indicate that there are two chain conformations [262] and the availability of a non-conjugated conformation may be a key to the low transition temperatures and solubility, when compared to the stiff-chain conjugated polymers.

7 Diffraction Structures and Chain Packing in the Crystal

The determination of crystal structures of polymers by X-ray or electron diffraction is much more difficult than the determination of structures of other organic crystals because large single crystals are not formed. Diffraction measurements on unoriented samples typically give only a few reflections, which may be used to suggest a unit cell for simple systems, but structures are normally derived from studies on highly oriented and annealed samples. Orientation will align the chain axes with the sample axis and so allow separation of the lattice reflections along the chain (c) direction from the others. Annealing will sharpen the reflections by increasing the crystal size and reducing the level of disorder.

The pioneering studies of polymer crystal structures were those of Natta [435] which followed from the availability of large numbers of stereoregular, crystalline polyolefins produced by the Ziegler catalysts. Based on his studies, Natta developed a set of rules governing polymer structures. He found that for most systems the helical conformation observed in the crystalline state was the same as the lowest energy helix expected for an isolated macromolecule, which is to say that only near-neighbour intrachain interactions are important. In order to pack as regular structures, the chains must form regular helices, in that any regular arrangement of points along a line can be described as a helix. In the polyolefins the helices then pack so as to most completely fill the available volume and so maximise the non-directional van der Waals interactions between the chains. This may involve neighbouring helices of the same or opposite twists and, if the chains also have a directionality, neighbours may be parallel or antiparallel. A measure of the goodness of packing of the crystal is the packing density, the fraction of the total volume which is filled by atoms. In some systems, such as the polyamides (Nylons), hydrogen bonding or other strong forces between the chains can dominate the interchain packing or modify the helical conformation.

Given that all polymer crystal structures show parallel chains, cubic unit cells do not occur. The simplest structure, which would be expected for smooth cylinders in a close-packed parallel array, is hexagonal with a packing density of 78%. Many simple polymers, including polyethylene, show an orthorhombic structure which is only slightly distorted from hexagonal. Wunderlich [436] gives a good general discussion of polymer structures, packing densities and unit cells.

In the case of the conducting polymers we would expect orbital delocalization along the chain to lead to a well-defined energy minimum and hence a single stable conformation. The packing between chains should also be important for conductivity as chain-chain orbital overlap will affect the electron transport between chains. This overlap should be discernible from chain-chain spacings. Any lowering of energy associated with highly directional interactions between neighbouring chains should also be reflected in highly anisotropic packing in the a–b plane of the chains as opposed to a simple close packing.

7.1 Polyacetylene

Acetylene polymerizes to the *cis*-configuration at low temperatures, presumably reflecting the geometry of the catalyst-monomer-polymer complex (trans-opening of double bonds is unknown in coordination polymerization and would require a very unusual geometry at the active centre). On heating to temperatures above 20 °C the *cis*-form isomerises to *trans*-polyacetylene. Two structures are possible for the *cis*-form (see Fig. 1), *cis-transoid* or *trans-cisoid*, depending on whether the double of the single bonds are in the *cis*-configuration. Calculations agree that the *cis*-form is less stable than *trans*-but there is more doubt over the energy difference between the two *cis*-forms [437, 438]. The *cis-cisoid* form is unknown.

Cis-polyacetylene shows 10 reflections in x-ray diffraction [439]. The unit cell was identified as orthorhombic, $a = 761$ pm, $b = 447$ pm, $c = 439$ pm with a density of 1.16 g cm^{-3}. In the original x-ray study the b axis was taken to be the chain axis but subsequent electron diffraction studies allowed fibre patterns to be obtained [440]. On the principle that, for all polymers, the fibre axis is the chain axis, c was identified as the chain direction although there is some dispute [24]. The analysis cannot definitely distinguish between the *cis-transoid* and *trans-cisoid* structures.

Bates and Baker [441] reported crystallization of *cis*-polyacetylene from solutions of a polyacetylene-polystyrene block copolymer and claimed that the unit cell was hexagonal with the chains packed in a helical conformation, having six carbon atoms per turn, a unit cell length (c) of 484 pm and an inter-chain spacing of 512 pm. Elert et al. [442, 443] have calculated the energetics of helix formation in cis-polyacetylene. They suggest that the helical conformation of the isolated chain is more stable than the planar form by about 2.5 kJ (mol CH)$^{-1}$, but that interchain forces in the crystal lattice favour the planar form by the same amount. They propose that both conformations can exist depending on the conditions of crystal formation.

For *trans*-polyacetylene, X-ray diffraction yields a monoclinic unit cell with $a = 424$, $b = 732$, c (chain axis) $= 246$ pm and $\beta = 91$ to 93°, i.e. very close to

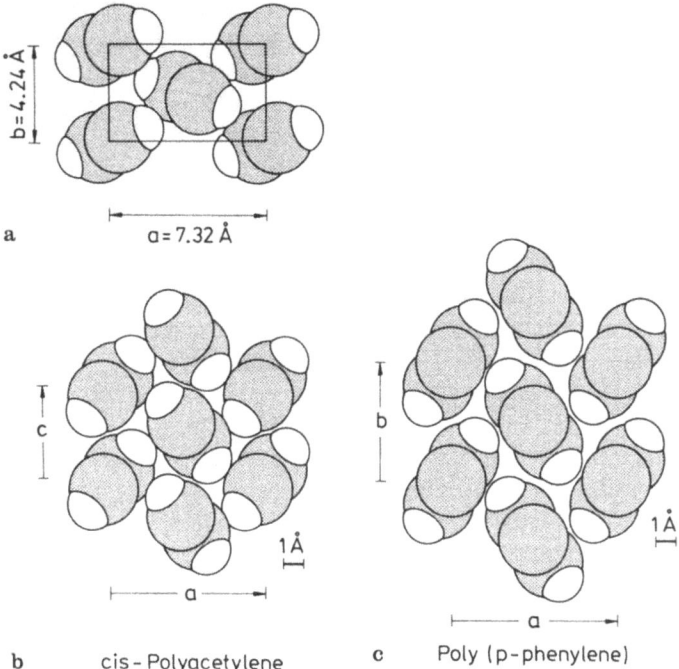

$b = 4.24$ Å

a

$a = 7.32$ Å

b   cis - Polyacetylene

c   Poly (p-phenylene)

1Å

1Å

Fig. 14a–c. Crystal structures of **a** *trans*-polyacetylene, **b** *cis*-polyacetylene and **c** polyphenylene. Reproduced with permission **a** from Ref. [6] p. 103, **b** and **c** from Ref. [286]

orthorhombic, which is in turn quite close to hexagonal. The setting angle, between the plane of the C–C bonds and the b axis in the ab plane is variously estimated to be 55° or 24° (Fig. 14).

Another parameter of considerable interest for understanding the behaviour of polyacetylene is the bond-length alternation. A simple model of perfect conjugation would lead one to expect equal lengths for the C=C and C—C bonds in trans-polyacetylene but this structure is unstable. On the basis of the intensity of the weak (001) reflection, Chien [6] argues that the bonds are 138 pm and 143 pm compared to normal double and single bonds of 135 pm and 154 pm. Fincher et al. [444] and Kahlert et al. [379] arrive at similar results from their x-ray data. In terms of polymer properties, this double-bond character in the single C—C bond certainly suggests a rigid polymer chain with little rotational motion possible. In deriving the structure of *cis*-polyacetylene Shacklette et al. [434] took the bond lengths to be 135 pm and 146 pm based on analogous small molecules.

The crystallinity of *cis*-polyacetylene has been estimated to be 76–84% by x-ray diffraction while *trans*-polyacetylene is 71–79% [6]. Observations on Durham polyacetylene [445] showed that the diffraction peak narrowed, and the interchain d-spacing decreased, during isomerization and annealing. The x-ray coherence length, a measure of the crystallite size perpendicular to the chains, increased from 2.6 nm to 7.1 nm, compared with 30 nm for polyethylene.

The isomerization of polyacetylene from the *cis*- to the *trans*-form can be readily followed by ir spectroscopy [347], by solid state NMR [446] or by X-ray diffraction [447, 448]. The transformation takes place without a significant loss of x-ray order and with a progressive change in the unit cell, which suggests that crystals in the partly transformed state contain either randomly [433] dispersed units in both *cis-* and *trans-*conformation or a mixture of *cis-* and *trans-*chains [432]. The isomerization is catalyzed by oxygen [447], and by dopants, and may involve the motion of a free radical along the chain [347].

The structures of the doped states of the conducting polymers are even less well established than those of the undoped materials since the x-ray patterns generally contain only a few broad lines attributable to the doped state. The best information on structural changes during oxidative doping of polyacetylene comes from studies on iodine doping. Baughman et al. [439] and Hsu et al. [449] found that doping produced an additional x-ray peak at 760–790 pm which they interpreted in terms of layers of iodine intercalated between the (100) planes of polyacetylene. This model has been considered in more detail by Shimamura et al. [450], who fitted the structure to observed electron diffraction intensities for eight reflections. Monkenbusch et al. [451] made similar measurements and concluded that the material went through a partly doped state which has both undoped polyacetylene, a partly doped amorphous state and then, at high levels, a fully doped regular structure. The onset of conductivity is attributed to percolation between these islands of highly-doped metallic iodine-polyacetylene complex. At low doping levels the dopant accelerates the *cis-trans* isomerisation.

The state of the iodine counter-ions has been studied by Mossbauer spectroscopy [452, 453] which shows the presence of both I_3^- and I_5^- with the latter predominating at high levels of iodine (23 mol% based on CH). This puts the maximum doping level at about 5 mol% counterions. Raman spectroscopy [454] and XPS confirm the presence of iodine polyanions. On the basis of intercalated layers of iodine packed as in the iodine crystal, the maximum doping level would be expected to be $(CHI_{0.43})_x$, whereas the maximum level observed is $(CHI_{0.31})_x$, the greater spacing being attributable to the repulsion of the anions. Studies with carefully controlled doping and undoping give evidence for an interplanar spacing of 1400 pm at low dopant levels which is attributable to a third stage (three layers of polyacetylene per iodine layer), but there was no evidence for a second stage [455]. Staging has not been seen with any other oxidative dopants. In contrast, doping of substituted polyacetylenes with I_2 leads to very much lower conductivities than for polyacetylene itself and the majority of the iodine is present as I_2, with a small proportion as I_3^- [456].

Staging is clearly observed during the doping of graphite [457], where a series of doped structures are observed with intercalated planes of counterions separated by varying numbers of planes of graphite. Such an effect would only be expected in polymers under conditions of slow, quasi-equilibrium doping. In graphite, it reflects the large energy needed to separate planes in order to allow in the first few ions. In the conducting polymers, the chain structures and the higher levels of disorder would be expected to make the appearance of staging less likely than in graphite.

Studies on AsF_5-doped polyacetylene show the counterion to be AsF_6^- [458]. On doping one mole of AsF_3 is released for every three moles of AsF_5 absorbed so that Mossbauer and x-ray spectroscopy during doping show signals assignable to AsF_5, AsF_6^- and AsF_3 [459]. Neutron diffraction studies on AsF_5-doped polyacetylene show a crystalline structure with crystal dimensions of the order of 10 nm and a similar structure to that found for iodine doping [460, 461].

A report on doping of polyacetylene by metal halides [462, 463] shows that the interplanar spacing increases with the size of the anion and clustering is inferred to occur at low dopant levels as the dopant reflection appears at about 3 mol% while much of the material is still undoped. It is not totally clear whether similar effects might be the result of a combination of slow dopant diffusion and a diffusion coefficient which is dependent on dopant concentration; this is discussed in more detail below.

Alkali-metal doped polyacetylene can also be viewed in terms of a layer structure [464]. The spacing of the planes goes from 450 pm for Li doping to 642 pm for Cs doping. Electrochemical doping with potassium [465] shows a series of doping stages with a spacing of 820 pm at 6 mol% and below, shifting to a spacing of 600 pm at 12% and above. Plots of voltage vs %dopant show one or two plateaux corresponding to stable phases of doping (Fig. 15). The fully doped "first stage" corresponds to 6 CH units per cation (16.7 mol%), with the cations in channels surrounded by four poly-acetylene chains (see Fig. 16). As cations are removed the channels slowly empty until a critical level is reached at which the structure rearranges to a second stage to leave fewer channels. A similar model is proposed for polyphenylene.

Shacklette et al. [464, 465] favour a structure in which the metal ions are intercalated in columns within the polyacetylene (Fig. 16). The argument against layer structures is based on the charge repulsion which would occur between adjacent metal ions. These structures also give agreement with the observed planar spacings [466, 467]. In the case of lithium and sodium the metal ion is small enough to fit within the polyacetylene structure with little distortion. For potassium there is a close fit of the metal inside the columns of polymer and hence a strong tendency to form a regular structure.

The random orientation of the crystalline order in typical Shirakawa polyacetylene means that diffraction studies are limited to powder methods. For such studies, and many others, it would be very useful to have much more oriented polymers and many attempts have been made to orient polyacetylene, either by mechanical treatment of the polymer or by appropriate modifications to the polymerization reaction. These have been reviewed earlier.

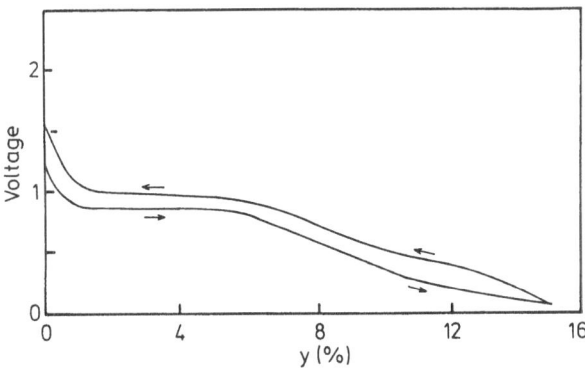

Fig. 15. Evidence for staging in doping of polyacetylene. Reproduced with permission from Ref. [465]

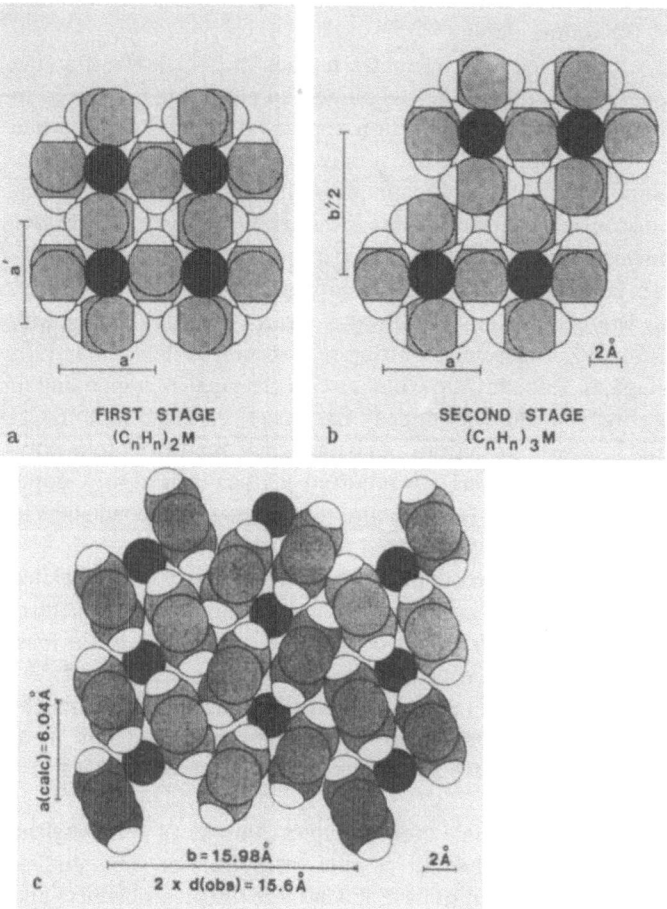

Fig. 16a–c. Proposed crystal structures for **a** metal-ion doped polyacetylenes and **b** metal-ion doped polyphenylene. Reproduced with permission from Ref. [465]

We have already described the particularly interesting property of Durham poly-acetylene that it can be stretched to draw ratios of up to 20 during the transformation, to yield a polyacetylene sample with high levels of orientation. Leising [380] drew single fibres of Durham polyacetylene to a highly oriented *trans*-state with a density of 1.06 g cm^{-3} and found a monoclinic unit cell of $a = 426$ pm, $b = 733$ pm, $c = 246$ pm and $\beta = 91.4°$. More recently, Kahlert et al. [468] have obtained a structure for *trans*-polyacetylene from very highly drawn fibres of Durham material. They give a monoclinic unit cell of $a = 418$ pm, $b = 734$ pm and $c = 245.5$ pm, $\beta = 90.5°$ and a setting angle of $61°$.

Friend et al. [469, 470] have reported X-ray diffraction studies of highly drawn films prepared from Durham polyacetylene. They analysed the width of the diffraction peaks to obtain a crystallite size perpendicular to the chains of 5 nm in Durham poly-acetylene compared to 10 nm in Shirakawa polyacetylene and distortion parameters within the crystals of 21 pm and 15 pm respectively. Moon et al. [471] suggest that anti-phase stacking of the bond alternation in the polyacetylene chains within the unit cell is the favoured packing in oriented samples. They found that Durham polyacetylene can be doped electrochemically up to 6 mol% Na^+ and will revert to its original structure on undoping, with only a slight reduction in coherence length. Doping to higher extents, with larger ions or with several doping/undoping cycles led to a significant reduction in crystallinity.

7.2 Polyphenylene

The structure and properties of poly(p-phenylene) have recently been reviewed by Elsenbaumer and Shacklette [386]. As normally prepared it is a poorly crystalline material with one strong, broad x-ray diffraction peak at 464 pm and four other clear lines, although the pattern sharpens on annealing. Consequently the crystal structure is not well established. Kovacic and co-workers [472] have reported an orthorhombic unit cell ($a = 781$ pm, $b = 553$ pm, $c = 420$ pm) in which the packing of chains is similar to that in cis-polyacetylene with the planes of adjacent chains perpendicular. They report that the observed diffraction pattern would also be consistent with a monoclinic unit cell, similar to that found for the crystalline oligomer sexiphenyl, with an angle $\beta = 100°$. On annealing at up to 400 °C the main (110) peak width decreased from 2.9 to 1.3 degrees 2θ while the intensity increased, and the d-spacing of the main peak decreased from 464 to 453 pm. Based on this unit cell, the bulk density is calculated as 1.39 g cm^{-3}, a packing density of 78%. The measured density of pressed powder was 1.25 g cm^{-3}.

From electron diffraction studies of fibres made using $MoCl_5$ as a catalyst, Teraoka [473] derived a monoclinic unit cell with $a = 779$ pm, $b = 562$ pm, $c = 426$ pm and $\beta = 79°$. Polyphenylene prepared using $FeCl_3$ as a catalyst showed only broad diffraction rings, suggesting a much less ordered form of the same structure. By neutron diffraction from deuterated polymer, Haslin and Riekel [474] found a monoclinic unit cell of $a = 781$ pm, $b = 553$ pm, $c = 426$ pm and $\beta = 100°$. Pradere et al. [475] found a similar structure by electron diffraction of fibrous polymer. In terms of understanding the conductivity, it is the extent of face to face packing between chains and the degree of conjugation along chains that are particularly interesting. From studies on oligomers, it is believed that rings are not parallel within the chains, but that successive rings are twisted by 23°, however there is still considerable orbital overlap along the chain.

The crystal size in polyphenylene, as determined from x-ray peak widths, is of the order of 5 nm [476] with a disorder parameter g = 0.026 nm. Compression at up to 12kB decreased the d-spacing perpendicular to the chains, decreased the peak size and increased the disorder slightly. Annealing at temperatures above 250 °C increases the crystal size and perfection [472]. The spin concentration increases above 300 °C, but unlike those in polyacetylene, these spins are not mobile [477]. The crystallinity has variously been estimated as 80% [327] and 20 to 30% [478]. It seems to depend on the catalyst used in the Kovacic method. Polyphenylene produced by the precursor route has a crystallinity from 60–80% dependent on the conversion conditions [252].

Hasslin et al. [460] found that poly-p-phenylene converts to a "raft" structure, on doping with AsF_5, with layers of dopant separated by two layers of polymer chains and the planes of the benzene rings all parallel to the planes of dopant. The repeat spacing across the layers is 1070 pm and 426 pm parallel to the chains. There was only one structure formed, corresponding to $(C_6H_4)_2AsF_5$. Baughman and co-workers [467] propose a model for potassium-doped polyphenylene which is similar, except that the polymer is tilted at a setting angle of 25° to the dopant planes (Fig. 16). The potassium atoms do not sit above the rings but above the C—C bonds joining the rings, which is somewhat unexpected, but may serve to stabilise the quinonoid structure. The structure is orthorhombic $a = 604$ pm, $b = 1598$ pm, $c = 435$ pm (chain axis) and the repeat spacing across the planes (020) is 799 pm. The polymer backbone may be twisted slightly. There is evidence, from electrochemical doping with lithium and potassium [465], for structures with three and four planes of chains between succeeding planes of dopant atoms. With lithium vapour doping [479] the c-axis reflection remains and the others show a broadening, suggesting that the metal occupies empty sites between the chains in the pristine polyphenylene lattice. X-ray studies on SbF_5-doped oligomers give a similar structure with a triclinic unit cell in which the molecules are stacked with interplanar spacings of 340 pm and the stacks are separated by layers containing dopant counterions and included solvent [480].

Thin films of polyphenylene prepared under conditions of high shear are formed with the chains in the film plane [481], this is possibly the result of aggregation of particles which are elongated in the chain direction. These films were red with an absorption maximum at 500 nm which shifted to 380 nm on doping with SbF_5 or on exposure to moisture. The authors suggest that the as-prepared polymer is doped with tetrachloroantimonate counterions. Fibres of polyphenylene have been prepared by extrusion of the powder at elevated temperature [482]. A modest level of orientation was obtained (half-width of (110) peak of 37°).

The kinetics of doping of Kovacic polymer by AsF_5 have been reported to involve a combination of adsorption on the powder surface and diffusion [483]. The diffusion coefficient into a bulk sample is estimated to be 3×10^{-6} cm^2 s^{-1} at room temperature [482], a very high value which probably reflects porosity in the sample. Considering diffusion into the individual particles gives a diffusion coefficient of 10^{-15} cm^2 s^{-1}. These workers observe diffraction patterns at intermediate dopant levels corresponding only to pure polymer and doped material suggesting that there are only two states, fully doped and undoped, with no staging. This is in agreement with structural studies mentioned above which do not find staging for oxidatively doped polymers.

7.3 Poly(Phenylene Sulfide)

Poly(phenylene sulfide) is not a true conducting polymer, in so far as it becomes conducting when doped with arsenic pentafluoride only due to a conversion to a planar ladder-structure. In the undoped form the polymer is 65% crystalline with rings orthogonal along the chain, and hence no conjugation although there is orbital overlap between the sulphur and phenyl rings [434].

7.4 Polypyrrole

The structure of polypyrrole has recently been reviewed by Street [393]. As normally formed by electrochemical polymerization, it is highly disordered, essentially an organic glass. Wide angle x-ray studies have shown anisotropic organization with respect to the electrode surface in electrochemically synthesised samples [484]. This high level of disorder is supported by XPS studies showing a broad range of C_{1s} energies [485, 486]. XPS also shows that polymerization is mainly through the α-positions but there is also some reaction at the β-position. A more perfect structure can be obtained by polymerization of 3,4-dimethylpyrrole where the β-positions are blocked. Molecular models suggest that pyrrole could form planar cycles of ten units or flat spirals [487]. However, Street et al. [488] favour a linear planar structure with the rings alternating either side of the backbone. Pyrrole dimers and trimers are linear and planar, with successive chains stacked parallel. Electron diffraction from undoped polypyrrole, fitted to the planar structure model, gives a monoclinic unit cell with a chain-chain spacing of 341 pm in polypyrrole and 365 pm in polydimethylpyrrole [489]. Since conjugation can be preserved with the rings alternating or on the same side of the chain, and rotation from one form to the other would involve loss of conjugation, it is unsurprising that crystallization is difficult in polypyrrole and polythiopene. Side groups which could force an alternating structure might enhance crystallinity.

The conductivity of doped poly(3,4-dimethylpyrrole) is 10 S cm^{-1}, while conductivities of poly(diphenylpyrrole) and of poly(N-methyl pyrrole) are both about $10^{-3} \text{ S cm}^{-1}$, very much reduced from pyrrole and dimethylpyrrole. This is attributed by Street to non-planarity of the substituted polymers [393].

XPS of BF_4^--doped polymer with four pyrrole rings per anion shows that all the pyrrole rings are equally charged. On electrochemical undoping and redoping in aqueous electrolytes, the new state has oxygen-containing counterions, presumably OH^-, which replace the tetrafluoroborate and are involved in degradation reactions (see Sect. 9.4).

Polypyrrole can be prepared with n-alkyl sulfates and sulfonatés as anions [490], forming layered structures with a bilayer of the detergent separating layers of polypyrrole.

7.5 Polythiophene

Studies on polythiophene have been reviewed by Tourillon [136]. Polythiophene prepared by electrochemical polymerization appears to be completely amorphous by X-ray diffraction. The film density is 1.4 to 1.6 g cm^{-3}. Mo et al. [491] found that polythiophene prepared by chemical coupling of diiodothiophene is semi-crystalline. The crystallinity, measured by x-ray diffraction, was around 35% for the as-prepared material and rose to around 56% on annealing in nitrogen at 380 °C. At the same time, the x-ray linewidths narrowed on annealing, suggesting an improvement in coherence length. The scattering was said to be consistent with either an orthorhombic unit cell, $a = 780$ pm, $b = 555$ pm, $c = 808$ pm, or a monoclinic cell, $a = 783$ pm, $b = 555$ pm, $c = 820$ pm, $\beta = 96°$. Bruckner et al. [492] have recently reported X-ray studies on a similar polymer, after annealing. They found the best fit to experimental data was by assuming coplanarity of the thiophene rings. They proposed a bidimensional model, with $a = 779$ pm and $b = 553$ pm and only partial ordering of the chains along the c-axis.

Poly(3-methylthiophene) doped with $CF_3SO_3^-$ has been found to have a small amount ($<5\%$) of highly crystalline material [493], which has a hexagonal diffraction pattern by transmission electron microscopy. The appearance of a small amount of highly crystalline material could be due to crystalline impurities or low molecular weight material. The unit cell has $a = 970$ pm and $c = 1220$ pm. Tourillon and Garnier [493] suggest that the structure has all thiophene units on the same side of the backbone (cis) with the chain wound into a spiral of 11 units with an outer diameter of 2 nm. Poly(3,4-dimethylthiophene) and poly(diethylthiophene) show a higher energy absorption maximum in the UV due to steric hindrance causing twisting of units along the chain and breaking the conjugation.

7.6 Poly(Phenylene-Vinylene)

Poly(phenylene-vinylene) is produced from a soluble precursor, by a route similar to that of Durham polyacetylene. Like the latter polymer it can be oriented by stretching the film or fibre during the process of transformation [494, 495]. Granier et al. [496] recently published a structural analysis based on electron diffraction from highly oriented films produced in this way. They propose a monoclinic unit cell containing two monomer units and with dimensions $c = 658$ pm, $a = 790$ pm, $b = 605$ nm and $\beta = 123°$. The molecules are said to be nearly perfectly aligned along the stretching direction but exhibit partial translational disorder. This conclusion is in agreement with the ir analysis of Bradley et al. [497], who suggested a very high degree of orientation, based on an analysis of the infrared dichroism of stretched films. Subsequently, x-ray diffraction has confirmed an orientation function of about 0.94 after drawing by $5\times$ and heat treating at 300 °C [498]. This degree of orientation is comparable to that in a material which has been drawn to $10\times$ the original length. Analysis of the peak widths yielded a crystallite size of 30 nm perpendicular to the chain axis.

Oriented material can be described in terms of a paracrystalline structure. On doping with AsF_5, order is retained, with a new spacing developing perpendicular to the chain axis and a regular positioning of dopant along the chains [499].

7.7 Discussion

Most of the conducting polymers have fundamentally regular chain structures and would be expected to crystallize like polyethylene, in contrast to the normally atactic polymers like polystyrene where chain irregularity prevents crystallization. X-ray studies of Shirakawa polyacetylene suggest that the structure may resemble polyethylene with a two-phase microstructure of amorphous regions and small crystallites. Most of the other polymers appear to be single phase with some degree of order. This is illustrated by our studies of Durham polyacetylene [445], where annealing converts the polymer from being clearly amorphous to largely crystalline but without any clear break corresponding to the onset of order. Rather, there seems to be a continuum from disorder to order (Fig. 17).

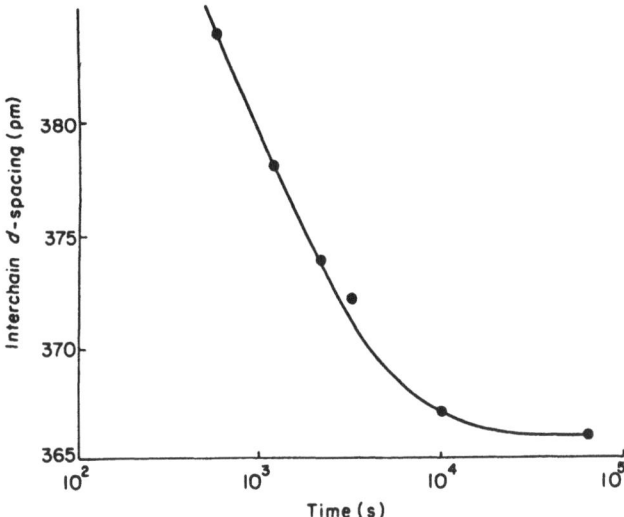

Fig. 17. Variation of inter-chain spacing in Durham polyacetylene with annealing time at 90 °C, showing increasing ordering. (Ref. [230])

Two sources for this behaviour suggest themselves. One, which could be applicable to polyphenylene, is that the material is a glassy mixture of short chains, where the varying chain lengths, coupled with low mobility, prevent extensive crystallization. Annealing at high temperatures then allows some fractionation to occur with an increase in crystallinity. The alternative view, which would be viable for Durham polyacetylene, is that the polymer chains are formed below the glass transition as random coils and never attain sufficient mobility to allow the extensive uncoiling necessary for crystallization. The removal of kinks from a chain requires diffusion of the kink along the whole length of the chain, and so the formation of long straight regions from a random coil requires liquid-like long range mobility. In the case of Shirakawa polyacetylene, the fact that the initial chain is formed in solvent allows considerable mobility during the short time before the nascent chain precipitates.

Polyacetylene has only one low energy conformation, that of the straight fully-*trans* conjugated chain. It is also very stiff, in that rotation about any bond is unlikely, in the absence of a conjugation-breaking defect such as a soliton (or free radical), since the activation energy barrier is large. The chain can thus be viewed as a series of short rods joined by hinges. The ordering of this structure must be quite different from the crystallization of polyethylene where internal rotation is very easy in the amorphous state. The x-ray observations on isomerization and drawing of the Durham polymer suggest that the starting state is essentially a nematic structure, with considerable local orientational order of the chains and this order progressively develops during annealing until the chains finally achieve both lateral order and registration parallel to the chain axes. Drawing extends the range of this ordering by biasing the average chain direction along one axis and so reducing the number of boundaries between mismatched domains. The behaviour of the rigid chain polyamides such as poly-*p*-benzamide and Kevlar is apparently similar [500]. It should be remembered that a high degree of orientation means that the chains are parallel but not necessarily straight, since an orientation function of 0.95 will allow roughly 5% of the units to be in hairpin bends where the chain reverses direction.

The soluble polythiophenes are the first conducting polymers that can be taken above their glass transition without decomposition and it will be interesting to study morphology-property relationships. Heeger et al. [262] have recently described conformational changes in solutions of poly-3-hexylthiophene which seem to involve a coil-helix transformation as the temperature is decreased or a poor solvent is added.

In general, doping tends to lead to a loss of x-ray order in polyacetylene and polyphenylene, suggesting that dopant ions may be distributed more or less at random. The structural models shown in Fig. 16 are clearly idealised as only limited order is seen even in cation-doped polymer. The anion dopants are much larger and apparently disrupt the structure too much for any sign of regularity to be seen, except in the case of iodine.

8 Transport Properties and Doping

The diffusion of dopants into, and out of, conducting polymers is important for possible applications in batteries, and as conductors or semiconductors. For conductors or semiconductors, the chief requirements are that the material can be doped in a reasonable time, but that it will then not lose dopant over periods of years. This is particularly important in determining junction stability in devices. In the case of batteries, on the other hand, rapid and reversible uptake and loss of dopants is needed, since the diffusion rate controls charging and discharging rates. In addition, the accessibility of the structure to oxygen, and other degradants, will be a factor in the stability of the polymer.

In many cases the important property will actually be the permeability, which is the product of the diffusion coefficient and the solubility of the dopant in the polymer. The solubility is determined by the degree of interaction between the diffusant and the polymer. The picture is further complicated, since reactions may take place so that several different species are diffusing. The reaction of a gaseous dopant with a conducting polymer is a complicated diffusion and reaction process. We must consider the solubility and diffusion of the molecular gas, the charge-transfer reaction to dope the polymer, the diffusion of the resulting ions in the doped (intercalated) structure and any reaction between the dopant ions and the polymer which may lead to covalent bonding.

Diffusion in crystalline solids can be divided into vacancy and interstitial diffusion. Large diffusants are presumed to occupy lattice sites and to move by jumping into adjacent vacancies. The activation energy for diffusion is then the sum of the jump energy and the vacancy creation energy. Small molecules can move by jumping between empty interstitial sites within the lattice, with a much smaller activation energy. Disordered materials do not show such clear types of sites, but the diffusion coefficient and activation energy will be a strong function of the size of the diffusant. As doping occurs "intercalation" may cause planes to open up within the structure, which allows easy access to further dopant, hence the diffusion coefficient will be a strong function of local dopant concentration.

Fickian diffusion of a solute into a continuous polymer film would be expected to give a weight-uptake curve which is initially linear with the square root of time and then levels off, but which shows no induction time [501]. A common response of polymers to swelling solvents, is Case II diffusion which can show an induction time but the weight uptake then increases linearly with time [502]. The diffusion of alcohols into polymethylmethacrylate has been recently described by Thomas and Windle [503]. In Case II, the rate of advance of the diffusant is controlled by the relaxation time of the polymer at the interface between a highly swollen region and dense polymer. The swelling of the polymer results in a greatly enhanced diffusion rate in the swollen region.

8.1 Polyacetylene

The diffusion behaviour of Shirakawa polyacetylene is complicated by its fibrillar morphology and high surface area, so that weight changes depend on pore transport and surface adsorption, as well as on diffusion into the fibrils. Chien [6] has reviewed earlier studies of the diffusion of dopant counter-ions in Shirakawa polymer and has emphasised the wide range of values of diffusion coefficient which are reported and which depend a great deal upon the morphological model chosen to interpret experimental data.

Danno et al. [504] have made diffusion measurements from the weight uptake of iodine by Shirakawa polyacetylene. They obtain values for the diffusion coefficient which vary from 4×10^{-10} to 1.4×10^{-9} cm^2 s^{-1} with increasing iodine pressure. Unfortunately, they base their calculations on the film thickness rather than the fibril diameter, and the true values are thus orders of magnitude lower. Similar assumptions were made by Beniere et al. [505], who obtained macroscopic diffusion coefficients in the range 10^{-10} to 10^{-8} cm^2 s^{-1} for Shirakawa polyacetylene. In contrast, Pekker et al. [506] estimated a value of 3×10^{-12} cm^2 s^{-1} for iodine diffusion in the fibrils, from microwave measurements of conductivity changes as the dopant diffused from the polymer into an encapsulating resin. Radici et al. [507] monitored the change of the ESR signal from lightly Li$^+$-doped polymer with time and derived an intra-fibrillar diffusion coefficient of 2×10^{-17} cm^2 s^{-1}. Kaner and MacDiarmid [508] have measured the intra-fibrillar diffusion coefficient for Li$^+$ as 4×10^{-18} cm^2 s^{-1}. Armand [509] reports values of 10^{-15}–10^{-18} cm^2 s^{-1} for Li$^+$ in polyacetylene. An increase of diffusion coefficient with applied field, "field enhanced diffusion", was suggested by MacDiarmid and co-workers but has been discounted by Armand [509].

Diffusion of dopants into conducting polymers can also be measured electrochemically [510]. The transport of ions into or out of a film, in an electrolyte solution, can be followed by passing a short current pulse to deposit an excess of doping ion into the surface of a film. By following the relaxation of the surface potential as the excess dopant diffuses in, the diffusion coefficient can be derived from a plot of potential versus (time)$^{1/2}$. The application of this method to Shirakawa polyacetylene has been described by Will [511], who derived a value of 5.7×10^{-12} cm^2 s^{-1} for the diffusion coefficient of BF$_4^-$ ions in the polymer fibrils and showed that the diffusion behaviour is crucial to applications of the polymer in battery electrodes. He also suggests that the wide variation in diffusion coefficients determined electrochemically may arise from difficulties in wetting the polymer fibrils with the electrolyte solution. Francois et al. [512] have studied the doping of Shirakawa films with K^+ and Li^+ counter-ions. They measure interfibrillar diffusion coefficients of the order of 8×10^{-6} cm^2 s^{-1}, and deduce that the intrafibrillar values must be of the order of 10^{-15} cm^2 s^{-1}. Chen et al. [513] showed that the macroscopic diffusion rate of iodine in Shirakawa polymer is reduced by a factor of 4 after compression to 10^4 bar, with a corresponding increase in resistance to oxidation.

Durham polyacetylene has the advantage of being a uniform, dense film and so lends itself much more readiliy to diffusion studies. In addition, the uniform morphology is much better suited to device applications, although the low surface area would limit applications in batteries. We have made extensive measurements on the doping of Durham *trans*-polyacetylene by gaseous AsF$_5$ [514, 515], which is believed to dope the polymer to form the hexafluoroarsenate ion and arsenic trifluoride [516].

$$3\,AsF_5 + 2(-CH=CH-)_x \rightarrow 2(-CH-CH^+-)_x + 2\,AsF_6^- + AsF_3$$

dopant diffuses in as a front, so that spectra at intermediate times show both fully doped and undoped regions.

As shown in Fig. 18, the weight increase of a film is roughly linear on a plot of weight versus (time)$^{1/2}$; in some cases there is a clear induction time. Analysis of the linear part of the plot yields a diffusion rate of 6×10^{-13} cm^2 s^{-1}. At long times the diffusion slows as the film saturates at 15 mol/100 CH. Esr studies suggest that the

Studies with AsF$_3$, which is not a doping agent, show that it is not taken up by undoped polyacetylene, but about 5 mol/100 CH dissolves into doped polyacetylene, from which the AsF$_3$ generated during the reaction, has been removed. Thus, neither AsF$_5$ nor AsF$_3$ are significantly soluble in polyacetylene but both are quite soluble in the doped material. An analogous observation is that doped polyphenylenesulphide is soluble in AsF$_3$ [258]. Both AsF$_3$ and AsF$_5$ show similar induction times for diffusion into Durham polyacetylene, suggesting that this is due to a skin effect on the films, arising from the transformation process [515]. An added complication is that the diffusion of AsF$_5$ is affected by the swelling caused by the AsF$_3$ that is produced during the reaction. We have recorded the diffusion of AsF$_3$ out of doped polyacetylene

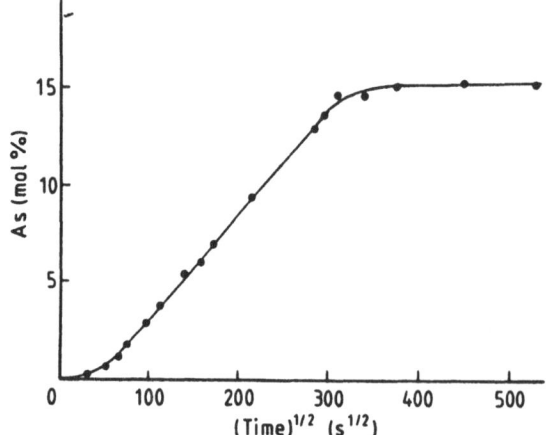

Fig. 18. Weight increase on exposure of a thin (3.5 μm) film of polyacetylene to AsF$_5$. (Ref. [515])

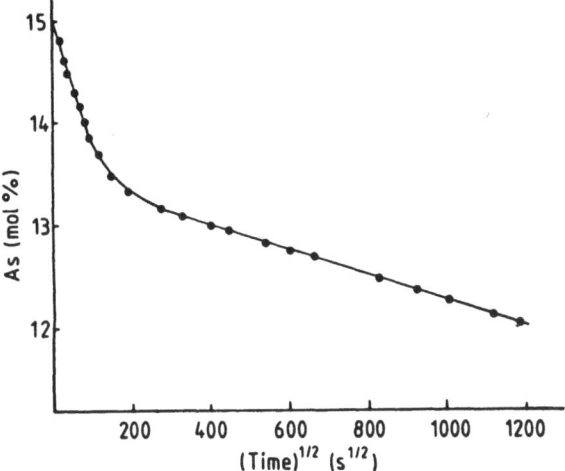

Fig. 19. Weight loss curve for diffusion of AsF_3 out of polyacetylene, showing the effect of deswelling of the surface in slowing diffusion. (Ref. [515])

and find a two-stage process, as illustrated in Fig. 19. The initial diffusion coefficient is 2.4×10^{-13} cm^2 s^{-1} but this decreases to 2.5×10^{-15} cm^2 s^{-1} as the surface becomes deswollen. If AsF_3 is allowed to diffuse in again, the measured diffusion coefficient is 2×10^{-13} cm^2 s^{-1}.

The diffusion coefficient for iodine, measured by weight uptake, is similar to that for AsF_5, 6×10^{-13} cm^2 s^{-1}. In contrast to AsF_5, a single esr line is seen and there is no induction time, suggesting that the doping process is much more uniform. The difference between AsF_5 and iodine may be due to a greater solubility of iodine in polyacetylene and a lower fully doped level, such that the contrast in permeability between the doped and undoped regions is much less than with AsF_5. It is possible that the doping reaction with iodine is intrinsically slower but first-order weight uptake kinetics would then be expected, rather than diffusion control.

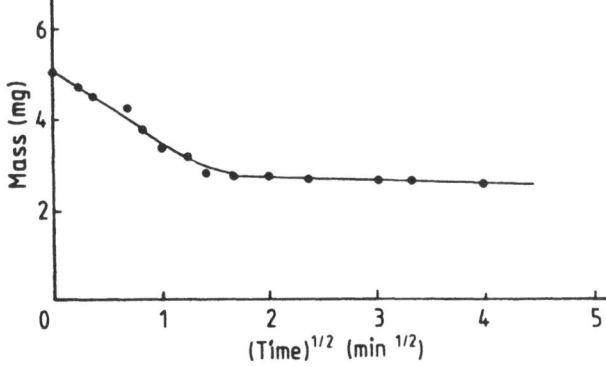

Fig. 20. Diffusion of a non-doping liquid out of polyacetylene. (Ref. [515])

A similar swelling effect to that observed with AsF_3 is seen for the loss of the fluoroxylene, formed during the initial decomposition of the precursor polymer to form predominantly *cis*-polyacetylene [515]. The diffusion coefficient is initially 4×10^{-9} cm^2 s^{-1} and decreases to 10^{-11} cm^2 s^{-1} as the polymer deswells, as shown in Fig. 20. In this case the swelling is not reversible.

Oxygen is also a dopant for polyacetylene, but on exposure the conductivity rises to a maximum then rapidly declines as oxidation of the polymer backbone occurs, as shown in Fig. 21. We have no data on the diffusion coefficient as the process is rapid and is masked by the reaction of oxygen with the polymer. The kinetics are first-order, implying that the doping reaction is rapid, goes to less than 1 mol%, and is then followed by irreversible oxidation of the polymer. Based on the observa-

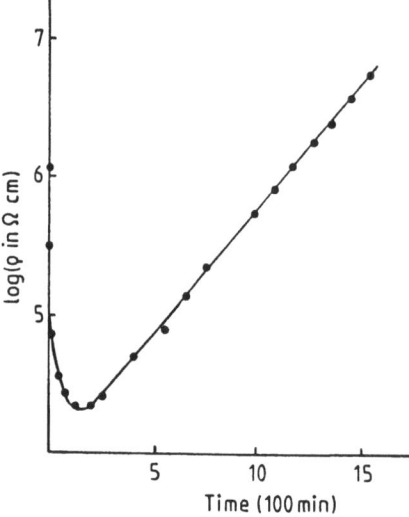

Fig. 21. Doping and subsequent degradation of polyacetylene by oxygen. (7 μm film, 50 °C, 600 Torr Oxygen) (Ref. [515])

Table 3. Diffusion coefficients for dopant counter ions

Polymer	Dopant	Diffusion coefficient cm^2 s^{-1}	Method
Polyacetylene	AsF_5	3×10^{-1}	Gas Phase [515]
Polyacetylene	I_2	6×10^{-13}	Gas Phase [515]
Polyacetylene	O_2	$>2 \times 10^{-12}$	Gas Phase [515]
Polyacetylene	ClO_4^-	1×10^{-14}	Electrochemical [515]
Polyacetylene	BF_4^-	8×10^{-14}	Electrochemical [515]
Polyacetylene	AsF_6^-	8×10^{-15}	Electrochemical [515]
Polyacetylene	Li^+	1×10^{-13}	Electrochemical [515]
Polyacetylene	Na^+	8×10^{-13}	Electrochemical [515]
Polythiophene	$(n\text{-Bu})_4 N^+ BF_4^-$	3×10^{-13}	Electrochemical [515]
Polythiophene	$Na^+ ClO_4^-$	6×10^{-12}	Electrochemical [515]
Polythiophene	$Li^+ AsF_6^-$	3×10^{-10}	Electrochemical [515]
Polythiophene	$Na^+ SbF_6^-$	1×10^{-15}	Electrochemical [515]
Polypyrrole	$Li^+ ClO_4^-$	5×10^{-12}	Electrochemical [501]
Polypyrrole	$Li^+ ClO_4^-$	5×10^{-10}	Electrochemical [518]

tion that a 7 μm film is fully doped in less than 300 min, we can estimate the diffusion coefficient for oxygen as being greater than 10^{-12} cm^2 s^{-1} at room temperature [515].

We have also estimated diffusion coefficients in Durham polymer using the electrochemical approach. Using this method the diffusion coefficient of Li$^+$ was found to be 2.5×10^{-13} cm^2 s^{-1} at a dopant level of 0.37 mol% and that of ClO$_4^-$ 1.3×10^{-14} cm^2 s^{-1} at a level of 0.05 mol%. Increasing the doping level caused the diffusion coefficient to fall slightly.

Table 3 summarises the diffusion coefficients measured for various gases and ions in Durham polyacetylene. The value for ClO$_4^-$ can be compared with the diffusion coefficient for AsF$_5$ doping discussed above. The higher value for gas doping can be attributed to the effect of AsF$_3$ as a swelling agent and possibly to the higher doping level in the gas-phase doped polymer. The slower diffusion of lithium when compared to the larger sodium ions, can be attributed to solvation of the lithium. For comparison, the diffusion coefficients of helium, oxygen and carbon dioxide in glassy polystyrene at 25 °C are 10^{-5}, 1.1×10^{-7} and 5.8×10^{-8} cm^2 s^{-1} respectively [517].

One concern with measurements of this type, is that undoping of a film may result from the outward diffusion of dopant ions or the inward diffusion of counterions which would then form salt within the film. This has been avoided in our polyacetylene study by measuring further doping pulses in samples which have only been doped in one direction, either reduction or oxidation.

8.2 Polypyrrole

By following the response of a film to a series of voltage steps, Mirebeau [518] obtained a value of 5×10^{-12} cm^2 s^{-1} for the diffusion coefficient of ClO$_4^-$ ions from lithium perchlorate in polypyrrole at 25 °C, with an activation energy of 30 kJ mol^{-1}. He found very similar values for undoping of either oxidized or reduced films, suggesting that the diffusant might be lithium ions rather than perchlorate. Genies and Pernaut [519] estimated diffusion coefficients for lithium perchlorate in polypyrrole and poly(N-methylpyrrole) as about 5×10^{-10} cm^2 s^{-1} and 4×10^{-9} cm^2 s^{-1} respectively. Their measurements were based on a colour change associated with doping.

Genies and Pernaut [519] also discuss the response of a doped film to an undoping potential. For a perchlorate-doped film being undoped in lithium perchlorate, the more mobile lithium ions first diffuse into the film to form lithium perchlorate, which then diffuses out slowly. Thus undoping studies of such a film would measure lithium diffusion, while increasing doping would measure perchlorate diffusion. This has been confirmed by use of a quartz microbalance which shows a weight increase on undoping polypyrrole perchlorate, due to the diffusion in of lithium ions [520].

These ion mobilities are very small, so that only very thin films can be reversibly switched between the doped and undoped states in a reasonable time scale. In contrast, application of polypyrrole as an electrode usually involves thicker films to prevent complete undoping during cycling.

Shinohara et al. [521] suggested that the diffusion rate of dopant ions in polypyrrole films depends upon the size of the anion used as counter-ion in the electrochemical synthesis so that films prepared with Cl$^-$ counter-ions were impermeable to larger ions, at least in the time-scale of a cyclic voltammetry experiment. Tietje-Girault et al. [522] made two-layer polypyrrole films in which a layer was prepared with a large counter-ion then a second layer, deposited on top with a much smaller counter-ion. They claimed that the resulting two-layer film is very much more resistant to leaching of the large counter-ion.

8.3 Polythiophene

The diffusion of a number of dopants in polythiophene has been followed by the electrochemical relaxation technique [515]. Polythiophene prepared from solutions of thiophene and tetrabutylammonium tetrafluoroborate in propylene carbonate, gave a diffusion coefficient at room temperature of 10^{-10} cm^2 s^{-1} and an activation energy for diffusion of 116 kJ mol^{-1}. This is a high value for activation energy but should be compared with diffusion of similarly sized species in glassy polymers [431], particulary as we expect a strong interaction between the chain and the counter-ion. The diffusion coefficient was also measured as a function of doping level from a mole ratio of 0.1 to 0.5 BF$_4$/thiophene unit; it decreases with increasing dopant ion concentration.

Similar measurements were also made on polythiophenes prepared from sodium perchlorate, lithium hexafluoroarsenate and sodium hexafluoroantimonate solutions. It was found that the diffusion coefficient first increased, then decreased, with the size of the anion, rather than simply decreasing as might be expected. This suggests that some of the smaller cation may also be incorporated into the film, possibly as dissolved salt, and then removed during the undoping step as discussed for polypyrrole by Genies and Pernaut [519]. This problem is not encountered in the polyacetylene measurements, since these films were not prepared electrochemically, and it should not influence the tetrafluoroborate results because of the large size of the counterion.

Kaneto et al. [523] have made measurements on the diffusion of lead perchlorate in polythiophene by following the colour change. They found a diffusion coefficient which varied from 10^{-10} to 10^{-12} cm^2 s^{-1}, depending upon the applied potential. The complexities introduced by morphological heterogeneity, counter-ion motion and solvent effects mean that further studies will be required to determine the relative importance of factors affecting diffusion in these materials.

The colour change from black to pale yellow, which accompanies undoping, has been proposed as the basis for optical switching devices [524], requiring very thin (less than 1 micron) films.

8.4 Discussion

There has been considerable interest in polyacetylene for rechargeable batteries as the high surface area of the fibrillar structure gives high current densities and the bulk density is low, so giving potentially high charge/weight ratios. Problems include slow degradation of the polymer by most dopants at high doping levels, and apparent irreversibility due to slow diffusion in the fibrils of Shirakawa polyacetylene. For semiconductor applications, diffusion of dopants must be very slow if homojunctions are to be stable between differently doped polymer layers. The diffusion rates measured so far, are far to high to allow for stable interfaces.

9 Stability and Degradation

When polymeric materials were first introduced on a large scale they were often thought to be chemically indestructible in normal use. This view arose because of the very high stability of low molecular weight models for the simple polymer chains but has proved true only for a very few polymer systems. There are two main reasons for the differences between polymers and their low molecular weight analogues. Firstly, polymers frequently contain defect structures capable of initiating degradation reactions at temperatures where the perfect chain would be unreactive. Secondly, the mechanical properties of a polymer depend on its high molecular weight and only a few chain scission reactions are needed to cause a rapid lowering of molecular weight. In practice the commercial development of many of the high-tonnage thermoplastics has depended almost entirely upon the understanding and control of their degradation reactions.

The degradation reactions of polymers have been widely reviewed [525]. In the absence of air, thermal reactions are the important degradation route. They may involve reactions of functional groups on the chain without chain scission, typified for example by the dehydrochlorination of PVC, or reactions involving chain scission, often followed by depropagation and chain-transfer reactions to yield complex mixtures of products. This latter route would be typical of the degradation of poly(methyl methacrylate), which depolymerizes smoothly to its monomer, and of polystyrene, which produces a wide range of tarry products.

Most polymers are actually used in air, where additional mechanisms of degradation are effective. Almost all synthetic polymers contain traces of peroxides, usually arising from reactions with oxygen during processing. These peroxides are susceptible to decomposition by heat, light or the catalytic effect of transition metal impurities. Their decomposition leads to radicals which can initiate autoaccelerating chain reactions with oxygen, producing more peroxides, other oxygen-containing products and chain scission. Photochemically-induced chain scission is the most important degradation process in polyolefins exposed outdoors and great efforts have been made to develop additives which can stabilize polymers by absorbing light, trapping free-radicals or decomposing peroxides. With the aid of appropriate additives the outdoor life of polypropylene film can be extended from less than six months to more than ten years.

Although we might expect similar principles to apply to the conducting polymers, an added dimension of these systems is the fact that they have electronic states of low ionization potential and high electron affinity ie. they are capable of acting as redox polymers and accepting or donating electrons. As discussed earlier, neither the ionization potential nor the oxidation potential are well defined for a typical conducting polymer, so that quantitative correlations are not possible.

In spite of the obvious importance of polymer stability in any potential applications of conducting polymers, there have been remarkably few systematic studies of degradation of polymers other than polyacetylene. Partly this may be due to an understandable reluctance of those involved in research on these materials to find that they are not stable and partly it is due to the difficulty of preparing samples in appropriate film forms for study. Another problem of discussing stability in conducting polymers is that there is no absolute standard for a stable material. For some applications an unchanging conductivity over a period of weeks might be regarded as stable whereas other applications might require stability over years. Thus when an author describes a material as a 'stable' conducting polymer, it is necessary to look carefully at the criterion of stability being used. In the following sections we review the present understanding of degradation mechanisms.

9.1 Undoped Polyacetylene

The thermal stability of polyacetylene in an inert atmosphere was investigated by Ito et al. [526], using DSC. They showed that the polymer is stable up to over 300 °C and identified benzene as the major product of thermal decomposition, with hydrogen, ethane, ethylene, propylene and butane as minor products of pyrolysis at 325 °C. The high-temperature pyrolysis has been reviewed by Chien [6] and studied in great detail by Chien et al. [527]. They found that the onset of rapid weight loss in a TGA experiment in helium was at 320 °C. By combined pyrolysis GC and mass spectroscopy they identified benzene as the major product of pyrolysis but also found substantial amounts of hydrogen-enriched products, such as saturated hydrocarbons and olefins. They propose that the initial thermal process is a random chain scission with the production of a terminal radical; intramolecular electron migration followed by cyclization will then yield benzene. In order to explain the formation of saturated products they postulate that electron-proton exchange processes are rapid. In our own studies of the isomerization of 'Durham' polyacetylene precursor we have found that the onset of significant degradation in a TGA experiment is at around 300 °C, slightly lower than for the typical Shirakawa material.

Pyrolysis of poly(methylacetylene) shows rather similar behaviour [528], with mesitylene as the major product but substantial yields of methyl and proton-enriched products. Thermal decomposition of this polymer sets in at around 150 °C and the mechanism is postulated to involve chain scission followed by cyclization reactions and both electron-proton and electron-methyl exchanges. Pyrolysis of poly(phenylacetylene) has been reported to start at 270 °C in nitrogen [529].

Although typical pyrolysis experiments show high onset temperatures for thermal degradation, this does not prove that degradation does not occur at lower temperatures. TGA, DSC and GC-MS methods all require that the polymer be degrading rapidly before reaction can be detected, whereas applications of polyacetylene might require stability at lower temperatures over very long time-scales. The long-term thermal stability of polyacetylene is an important problem which remains virtually unstudied. Druy et al. [530] showed that the dc conductivity of p-doped polyacetylene decays over periods of hours — days at 110 °C in vacuum. They interpreted the kinetics of this process as including a dopant-independent reaction, suggesting that dc conductivity may be very sensitive to small amounts of degradation. In studying polymers prepared via the Durham precursor route we find quite rapid decay in conductivity in vacuum at temperatures above 100 °C, with ir evidence for the formation of sp^3 carbon atoms, possibly via slow cross-linking. ESR spectroscopy also shows some loss of free spins under these conditions. It seems reasonable to conclude that the properties of polyacetylene are quite resistant to thermal degradation but long-term (years) storage at quite modestly elevated temperatures might lead to problems with stability of electrical properties.

Any possible sensitivity of polyacetylene to thermal degradation is overshadowed by its sensitivity to oxygen, which was detected at an early stage in studies of the polymer [531]. In their very early work Berets and Smith [1] showed that oxygen initially acts as a dopant for polyacetylene, with a corresponding increase in conductivity, but that the conductivity decreases again on further exposure. Shirakawa and Ikeda [532] reported that carbonyl bands appear in the ir spectrum of polyacetylene films on exposure to air. Another early observation on polyacetylene which caused great interest was the fact the film polymer, as normally prepared on the Shirakawa catalyst, contains a significant concentration of unpaired electrons [533, 534] as discussed earlier. Although solid state physicists have discussed the nature of these spins in terms of soliton models, in chemical terms they are delocalized free-radicals and would be expected to react easily with triplet oxygen. Snow et al. [535] found that the esr spectrum of polyacetylene is very sensitive to oxygen, which produces an increase in spin signal and a narrowing of the line width. They also showed that this effect is partially reversible on evacuation. Similar observations were reported by Vansco and Rockenbauer [536] and by Chien et al. [537], who suggested that oxygen acts both as a dopant and as a catalyst for *cis-trans* isomerization.

The effect of oxygen on polyacetylene was first systematically studied by Gibson and Pochan and their and other contributions have recently been reviewed [538]. These authors showed [539] that initial exposure to oxygen or air causes an increase of several orders of magnitude in the conductivity of *cis-* and *trans*-polyacetylene and that the effect is at least partially reversible on argon exposure. However, longer exposure (above 1–2 hr at room temperature) resulted in a rapid, irreversible decline in conductivity. These results were confirmed by dielectric studies [540] and it was suggested that oxygen is able to form a charge-transfer complex with polyacetylene in a manner similar to the reactions of conventional dopant species.

The first systematic kinetic study of the reaction of polyacetylene with oxygen was that of Yen et al. [541], who considered the possibility of using *trans*-polyacetylene as a thermal control coating for spacecraft applications. Using FTIR they showed that there are considerable structural changes in the polymer over periods up to 48 hr at 80 °C in dry air. The intensity of the $C=C-H$ bending and stretching modes decreased with time of exposure, accompanied by the appearance of a broad band around 3400 cm^{-1}, attributed to peroxides, and overlapping bands around 1700 cm^{-1}, attributed to mono- and di-ketones. Kinetic analysis of the changes in the $C=C-H$ stretch at 3200 cm^{-1} showed first-order kinetics with a half-life of 14.7 hr at 80 °C and an activation energy of 58 kJ mol^{-1}. This value is much lower than typical activation energies for polymer oxidation, presumably reflecting the facile complexation of oxygen by the polymer. Helmie et al. [542] also obtained evidence for oxygenated species in air-exposed polyacetylene by ^{13}C CPMAS nmr spectroscopy and Sohma et al. [543] used the same technique to show the formation of sp^3 cross-links during oxidation. Higashimura et al. [544] showed that the degradation of polyacetylene is greatly accelerated by gamma irradiation. A more extensive kinetic study was reported by Pochan et al. [545] who measured the effects of oxygen on the conductivity of polyacetylene samples over the range 50 to 110 °C. They again found an initial doping reaction, with the conductivity reaching its maximum value at 1 molecule of oxygen per 7 repeat units of the polymer, then decreasing rapidly. The decay in conductivity gave first-order plots which had two straight line sections, corresponding to an initial rate constant about 50% smaller than that observed later. The relative extent of these two regions depended on the *cis*-content of the polymer; high *cis*-materials showed only the larger of the two rates, whilst both were observed for high-*trans*-materials. The activation energies for the two pseudo-rate-constants are identical for a given initial *trans*-content but increase with increasing *trans*-content, to a value of 58 kJ mol^{-1} for 98% *trans*-polymer. The authors proposed that these data could be reconciled by a morphological model involving initial oxidation of the inter-fibrillar contact points. However, there is a real problem in interpreting data of this type since the relation of conductivity to chemical structure in these partially degraded polymers is not known.

In a later study [546] Gibson and Pochan used chemical methods to study the oxidation of polyacetylene, following the reaction over periods of thousands of hours at 25 °C. They observed the appearance of ir bands attributed to saturated and unsaturated ketones, and to hydroxyl groups, together with weaker bands assigned to peroxides or epoxides. At the same time the bands due to the polyacetylene chain decreased in intensity. As measured by oxygen content or ir analysis the reaction showed a short (22 h) induction period at 25 °C, which was not observed in weight uptake data. It was suggested that the difference arises from the fact that the oxygen is initially absorbed as the charge-transfer complex, which is not detected by ir or oxygen analysis. The subsequent degradation showed a good fit to first-order kinetics, although break points were again observed in the first-order plots (Fig. 22). This result is important because it shows that the degradation is not controlled by the rate of oxygen diffusion, which would lead to a $t^{1/2}$ dependence. The break points were again attributed to a two-stage oxidation reaction, involving an initial attack on the surfaces of the fibrils, with damage to the inter-fibrillar contacts, followed by a slower attack on the bulk material.

The possible involvement of oxygen in *cis-trans* isomerisation of polyacetylene and the related question of whether *cis*- and *trans*-polymers oxidize at different rates

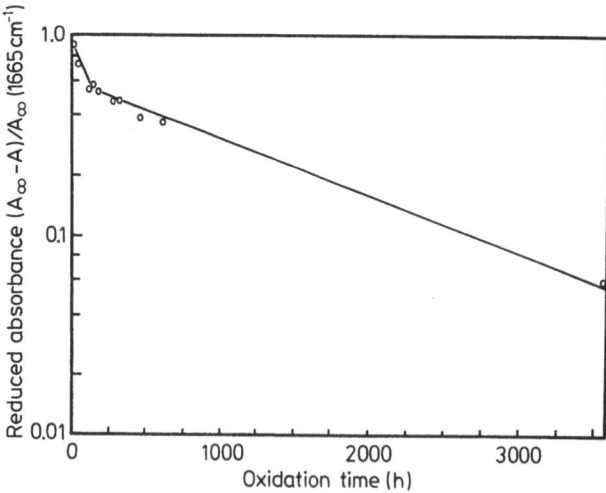

Fig. 22. Reduced absorbance of polyacetylene at 1665 cm^{-1} during oxidation at 25 °C. (Reproduced with permission from Ref. [546])

in the absence of simultaneous oxidation have proved to be contentious. In their studies of room temperature oxidation, Pochan and Gibson monitored the *cis-trans*-isomerisation by IR. They concluded that the oxidation does not induce isomerisation if the sample is kept in the dark as the rate of isomerisation at 25 °C was the same in both air and argon. They also suggested that the *cis*-isomer is selectively oxidized more rapidly than the *trans*- in ambient UV light, although a very early report [547] suggested that exposure to UV does not induce isomerisation.

In contrast, Bernier et al. have produced both ESR [548] and Raman [549] evidence to suggest that *cis-trans* isomerisation is catalysed by oxygen exposure at room temperature. Chien and Yang [550] found that the isomerisation rate at 60 °C is 12 times faster in pure oxygen than in nitrogen, with an activation energy of 62.3 kJ mol^{-1}, apparently independent of *cis*-content. This contrasts with the purely thermal isomerisation which has an activation energy of 58.5 kJ mol^{-1} for high-*cis*-polymer, rising to 159 kJ mol^{-1} at high *trans*-contents [551]. They suggest that the swelling effect of the oxygen doping may reduce the activation energy required to rearrange the crystal structure by thermal isomerisation, but do not specify at what level of exposure to ambient light their experiments were performed. Zanobi and d'Ilario [552] studied the reactions of polyacetylene with oxygen at 0 °C, monitoring by IR. They found that the *cis*-content of the polymer is virtually unaltered over 100 h in vacuum but more than 50% isomerisation occurs in the same period in oxygen. At the same time the rate of formation of carbonyl groups was about 3 times faster in an all-*cis*-polymer than in an all *trans*-material at the same temperature. They do not explicitly consider the interdependence of these two observations.

Cernia et al. [553] examined the effect of exposure on *cis-trans*-isomerisation, using a high intensity mercury arc lamp with no filtering specified. Under these aggressive conditions irradiation at minus 196 °C produced a rapid increase in free-spin concentration and irradiation at temperatures up to 120 °C increased the rate of isomerisation; above this temperature the thermal isomerisation is the dominant reaction.

In our own studies of the isomerisation of the 75% *cis*-polyacetylene produced by the Durham route [347] we have been unable to detect any effect on the isomerisation process of illumination with modest levels of light. We do find that the isomerisation is markedly affected by even trace amounts of oxygen, which lead to a change in the apparent order of reaction and a marked lowering of activation energy, somewhat similar to the observations of Chien and Yang [447] for conventional polymer. However, polymers prepared by the Durham route are very different from Shirakawa polymer and it would be unwise to extrapolate our results to Shirakawa materials.

To try to avoid possible problems associated with simultaneous *cis-trans* isomerisation and oxidation, Haleem et al. [554] studied all-*trans*-polymer, prepared by thermal isomerisation at 150 °C for 30 min. They measured resistance changes as a function of time at temperatures in the range 70–110 °C. After the initial doping reaction they found simple first-order kinetics, with no break points and an activation energy of 54.9 kJ mol^{-1}. In an extension of this study [555] it was suggested that the rate of doping is modestly increased by the presence of water vapour, which had very little effect on the rate of degradation. Exposure to ozone/oxygen mixtures (6 mol% O_3) also produced a higher rate of degradation. Huq and Farrington [556] also observed the sequential doping and degradation reactions and suggested that degradation is more rapid in moist air.

Conductivity changes in *trans*-polyacetylene have also been measured by Ebisawa and Tabei [557], who again saw sequential doping and degradation. They found first-order kinetics for the decay in conductivity, with no break points. Double logarithmic plots of conductivity versus temperature for the oxidized samples were linear, with slopes corresponding to an approximately $T^{1/4}$ dependence, consistent with the Kivelson inter-soliton hopping model for the conductivity [558]. It was shown that the first-order dependence of conductivity on oxidation is consistent with a model involving inter-soliton hopping together with the random introduction of conjugation-breaking defects.

Will and McKee [559] have described a detailed study of the oxidation of *trans*-polyacetylene powders in dry air over a wide range of temperatures from 25 to 142 °C. They followed the reaction by weight gain measurement and by photoacoustic ir measurements. Only at the highest temperature was there any sign of a limiting weight increase, corresponding to a 24% increase in sample weight. At 25 °C the initial rate of absorption was $0.04\%h^{-1}$, with the weight increasing by 40% in 2000 h and not stabilizing. The oxidation showed first-order kinetics at room temperature but there were deviations at higher temperatures due to simultaneous weight loss by evolution of volatile products. When corrected for this effect the data gave good first-order kinetics up to 90 °C, with an activation energy of 68.2 kJ mol^{-1}; above 90 °C there were deviations from the linear Arrhenius plot, indicating a change in mechanism. Ir studies showed a number of bands associated with oxidation products of the polymer, notably ketones and hydroxyl groups. There also bands at 890 and 1378 cm^{-1} associated with dopant-activated modes of the polymer chain. The functional group bands increased in intensity with the exposure time, whereas the dopant bands passed through a maximum intensity. It was concluded that a significant part of the oxygen is absorbed by the polymer as intercalated dopant counter-ions, which subsequently react with the polymer to induce degradation.

Chien et al. [560] studied oxidation of poly(methylacetylene) as a model for polyacetylene. This polymer differs from polyacetylene in that it is a wide band-gap insulator which is non-conducting, contains no free spins and is highly resistant to doping but is amorphous and soluble in organic solvents so that it can be cast into uniform films and additives incorporated. The absorption of oxygen began immediately on exposure of the samples, there was no induction period and the reaction followed first-order kinetics with an activation energy estimated as 61 kJ mol^{-1}. The oxidized polymer was soluble and its molecular weight and distribution were unchanged, implying no cross-linking or chain scission. Conventional stabilizers of polyolefin oxidation were modestly effective in stabilizing the polymer. The oxidation was photo-catalysed and the photo-oxidation difficult to inhibit. Yang and Chien [561] extended these studies to the oxidation of Shirakawa polyacetylene films, using volumetric and IR methods. At 70 °C the rate of oxygen absorption was about ten times slower than into poly(methylacetylene), presumably reflecting the high crystallinity of polyacetylene. However, the activation energy was much lower at 42 kJ mol^{-1}. Although some antioxidants were modestly effective in reducing the rate of oxidation of the polymer, most were ineffective and a spin-trapping additive (*N-t*-butyl-2-phenylnitrone), which is a highly effective stabilizer in polyolefins, was a pro-degradant in polyacetylene. It was not possible to establish the extent to which the failure to stabilize the polymer reflects the chemistry of the oxidation reaction or the difficulty of getting the stabilizer molecules into the lattice of the highly crystalline polymer.

The effect of oxygen exposure on the mechanical properties of polyacetylene was first described by Druy et al. [562], who showed that oxygen exposure reduces the elongation at break and makes the samples more brittle; they attribute this effect to cross-linking of the sample. More recently the viscoelastic properties of polyacetylene have been studied by Chen and Li [563]. They measured dynamic mechanical properties in a Rheovibron at 110 Hz, combining measurements of conductivity with measurements of tan δ and storage modulus. For *cis*-rich polymer samples the conductivity showed the expected maximum due to doping but tan δ and E showed complex behaviour. It was claimed to be possible to detect the four stages of reaction discussed earlier, viz: 1) oxygen-induced isomerisation to a higher *trans*-content, 2) oxygen doping via charge transfer, 3) reaction of oxygen with the polymer chains to give functional products and 4) cross-linking via the coupling reactions of radicals produced in the oxidation. For high *trans*-content samples the behaviour was said to be rather different, tan δ and E remaining rather constant over the same time period, whilst the conductivity falls rapidly; it was suggested that the rate of oxidation is much lower than that of the *cis*-rich polymer and that the rate of the cross-linking reaction (which consumes radicals) is higher than the rate of reaction with oxygen to produce new radicals so that the steady concentration of radicals is small.

There have been few studies of the effect of oxidation on the dopability of polyacetylene. Since the first stage of oxidation is the removal of the electrons of lowest ionization potential to form the charge-transfer complex, one would certainly expect oxidation to have some effect on the subsequent doping process. As far as we know, the only systematic studies are those of Pochan et al. [564], which have been reviewed [538]. These authors used samples oxidized at various levels, up to $(C_2H_2O_{0.55})_x$. The samples were allowed to react for the appropriate time then pumped to remove adsorbed oxygen and their response to iodine doping was monitored. In all cases the doping was effective in increasing the conductivity and at lower levels of oxidation the limiting conductivity was similar to that of the unoxidized material. At the highest oxidation levels the achievable conductivity was significantly reduced (from 10 to 10^{-5} S cm^{-1}). These results show that polyacetylene can survive significant amounts of degradation and remain dopable, confirming that some level of interruption of the conjugated sequences is quite tolerable.

9.2 Doped Polyacetylene

We have seen that the key factor in the degradation of polyacetylene by oxygen is the ability of oxygen to form charge-transfer complexes with the polymer. This gives a very high solubility for oxygen and introduces it into the polymer as the reactive superoxide anion. The ability to dope the polymer arises because the electron affinity of the O_2 molecule is high enough to allow it to remove the highest energy electrons from the polyacetylene band structure whilst the subsequent degradation arises because the polymer is reactive towards the superoxide ion. Nothing in this scheme is unique to oxygen. It might well be expected that other *p*-type dopants, of higher electron affinity, would compete effectively with oxygen for the electrons of lowest ionization potential, leading to stabilization of the polymer towards oxygen. Conversely, *n*-type dopants introduce new electrons into the polymer chain and might be expected to destabilize the polymer towards reaction with oxygen. At the same time it is reasonable to expect new chemical reactions between the dopant counter-ions and the polymer to introduce new possibilities for degradation.

In their studies of effects of oxidation of polyacetylene on its dopability, Pochan et al. [545] reported that iodine-doped polymer loses its conductivity in vacuum and concluded that the I_3^- counter-ions are able to react with the polymer chain, leading to iodination. Huq and Farrington [556] found that bromine- and iodine-doped poly-acetylenes lose conductivity rapidly at temperatures below 60 °C, whereas samples doped with AsF_5 are very much more stable.

Druy et al. [562] showed that iodine- and perchlorate-doped samples lose con-ductivity quite rapidly in vacuum, due to reaction of the polymer with the counter-ion. Yang and Chien [561] also observed the instability of these doped polymers and showed that the reaction of polyacetylene with perchlorate counter-ions can be explosive. They showed that doped samples are much more stable to oxygen than is the undoped material. Muller et al. [565] also observed that the stability of polyacetylene in air depends on the extent of doping, as did Ohtsuka et al. [566]. Aldissi [567] has suggested that iodine doped polyacetylene can be stabilized by phenolic antioxidants, although the effect was modest.

In our own work on Durham polyacetylene [568] we find that the stability of doped polymers depends upon the extent of doping. Thus when AsF_6^- is the counter-ion, a polymer doped to low levels (<1 mol%) shows very little change in conductivity over a period of days at room temperature in vacuum or dry air, whereas saturation doping (to about 17 mol%) produces a polymer whose conductivity decays rapidly, with ir evidence for the formation of C—F bonds in the polymer.

Pron et al. [569] looked at polyacetylene treated from the gas phase with H_2SO_4 which leads to HSO_4^- counter-ions. They found that the conductivity drops in air with the appearance of C=O bands in the ir, although the rate of decay is much lower than would be expected for undoped samples. The polymer was more rapidly degraded by exposure to water but could be redoped with further acid treatment. Pron et al. [570] have also reported hydrolytic instability in polyacetylene with $AlCl_4^-$ as the counter-ion. In both cases the proposed mechanism involves addition of OH^- to the chain and keto-enol tautomerism to form carbonyl groups.

In summary, the oxidation of polyacetylene has the following features: a) charge-transfer interaction with oxygen leads to doping, with an increase in conductivity, b) reaction of the polymer with the superoxide counter-ions leads to degradation with the introduction of oxygen-containing functional groups and loss in conductivity, c) doping with stronger electron acceptors removes the electrons responsible for oxygen doping and stabilizes the polymer to oxygen and d) low levels of non-reactive counter-ions can often be tolerated but higher concentrations or more reactive counter-ions lead to degradation with the introduction of new functional groups.

Chien [6] has pointed out that doping not only stabilizes polyacetylene towards oxidation but also stabilizes the dopant. The most obvious example is AsF_5, which reacts violently with atmospheric moisture but is stabilized in polyacetylene as the AsF_6^- counter-ion.

All of the studies so far discussed have involved gas-phase reactions. Recently, MacDiarmid et al. [571, 572] suggested that the same principles can be applied to poly-acetylene in contact with aqueous solutions. Since O_2^- is a hard base it would be expected to react preferentially with H^+ rather than with the polyacetylene cation. According to MacDiarmid, if polyacetylene is allowed to form its charge-transfer complex with oxygen then placed in contact with a solution of HBF_4, the oxygen is removed and replaced with BF_4^- counter-ions, leading to a polymer stable to further degradation. Similar results have been reported for other counter-ions [573, 574]. Terlemeyzan et al. [575] have tested these claims by ir studies. They were able to detect irreversible degradation with formation of carbonyl groups both in water and in aqueous HBF_4 solutions, although the rate of deterioration was somewhat slowed by the presence of the acid. It was suggested that the difference between these observations and the claims of MacDiarmid is due to the lower concentration of acid used in the ir studies. Since application of polyacetylene in battery or electrochemical applications will require cycling of the counter-ion concentration this is an important point, worthy of further study. Mu [576] and Will [577] have both studied degradation of polyacetylene in battery electrode applications and point out the need to avoid overdoping of the electrodes which leads to irreversible reactions with the dopant counter-ions.

If the principles, so far outlined, are valid then it is to be expected that n-type doping of polyacetylene would lead to a decrease in stability towards oxidation, and this is indeed so [578]. However, the introduction of electrons into the chain can also give a new instability in that the oxidation potential can fall to the point where the polymer is able to reduce water and it becomes hydrolytically unstable. Thus n-type doped polyacetylene reacts rapidly with water and with alcohols, with partial hydrogenation of the chain and a rapid decrease in conductivity [579, 580, 581]. Whitney and Wnek [582] have used the reaction of n-doped polyacetylene with alkyl halides and other reagents to prepare functionalized polyacetylene films.

Despite the various claims which have been made for the preparation of 'stable' polyacetylenes it is worth emphasising that stability is a relative property. It is our opinion that there has been no clear demonstration of a polyacetylene stable in air at even modestly elevated temperatures for more than a few days and that stability remains the Achilles heel of this particular polymer.

9.3 Polyacetylene-Related Polymers

As discussed earlier, substitution onto the polyacetylene chain invariably has a deleterious effect on dopability and conduction properties. At the same time the stability tends to improve. Masuda et al. [583] studied a large range of substituted polyacetylenes and found that stability increased with the number and bulkiness of the substituents, so that the polymers of aromatic disubstituted acetylenes were very stable, showing no reaction with air after 20 h at 160 °C. Unfortunately, none of these polymers is conducting. Deitz et al. [584] studied copolymers of acetylene and phenylacetylene; they found that poly(phenylacetylene) degrades even more rapidly than does polyacetylene and that the behaviour of copolymers is intermediate. Encapsulation of the iodine-doped polymers had little effect on the degradation, which is presumably at least in part due to iodination of the chain.

Pochan et al. [585] have studied poly(1,6-heptadiyne), a polyacetylene analogue, which forms dense continuous films. They find oxygen doping follwed by degradation in a manner similar to polyacetylene, except that the rate of degradation is much larger.

Since the polymer is readily swollen in aromatic solvents it is possible to introduce stabilizing additives and the effect of a spin-trapping antioxidant, based on tetramethyl piperidine, was studied [550]. It was found that the spin trap reduces the spin concentration in the polymer, lowers the conductivity and has a small stabilizing effect. They suggest that the piperidine acts as a reducing agent, coupling hydrogen atoms to the unpaired spins in the polymer and becoming converted to the nitrogen-centered radical. This is unlikely to be the whole story, since the piperidine radical is in itself highly reactive towards oxygen and Yang and Chien [561] showed that the derived nitroxide is a prodegradant for polyacetylene.

Aldissi [586] has reviewed the stability of block copolymers of acetylene and other non-conducting blocks. The copolymers are somewhat more stable than polyacetylene but the effect is small.

9.4 Other Conducting Polymers

We have shown that the overall instability of polyacetylene arises from two factors. The ionization (oxidation) potential of the polymer is low enough to allow it to solubilize oxygen in a charge-transfer complex and the chain is sufficiently reactive that it will react both with superoxide ions and with counter-ions introduced during doping. By analogy with polyacetylene we might expect to find that other undoped conducting polymers would be rather oxygen-sensitive, provided that their ionization potentials are low enough to allow charge-transfer interaction with oxygen. p-doped samples would be expected, however, to be stable to oxygen but similarly vulnerable to degradative reactions with their counter-ions.

The ionization potential of polyacetylene is about 4.7 eV, whereas that of undoped polypyrrole is rather lower at around 4.0 eV. We might therefore expect that polypyrrole would be similarly affected by oxygen. As prepared by electrochemical methods it is fully doped, so that the 'undoped' material results from electrochemical removal of the counter-ions, a process which is rarely complete and probably gives a non-equilibrium morphology. Pfluger et al. [587] showed that undoped polypyrrole absorbs large amounts of oxygen with an increase in conductivity from 10^{-6} to 10^{-2} S cm^{-1} occurring after absorption of 1 wt %; further absorption led to no change in conductivity. The oxygen uptake was irreversible. XPS studies showed chemical reactions at the nitrogen atoms but the changes in XPS spectra on oxidation could be reversed electrochemically. It was suggested that oxidation occurs at about one in every three nitrogen atoms, producing an unidentified counter-ion which can be removed electrochemically. It was also found that the same changes in XPS spectra occur on repeated electrochemical cycling of polypyrrole, suggesting that the original counter-ion is slowly replaced by an oxygen-derived ion. This result has been confirmed by Erlandsson et al. [588].

There is no doubt that doped polypyrrole is very much more stable than is polyacetylene, but reports are variable of exactly how stable it is. Street [393] reported that the polymer loses less than 20% of its initial conductivity after one year in air at ambient temperature and Diaz and Kanazawa [589] claimed that the polymer is stable at 100–200 °C, depending on the counter-ion; the latter authors also reported that polypyrrole is undoped reversibly by ammonia treatment. Munstedt [590] found that the conductivity of doped polypyrrole was unchanged after 200 days at 80 °C in argon, but drops rapidly at the same temperature in air. The rate of conductivity decay in air was sensitive to the counter-ion, suggesting that both the counter-ion and oxygen (or moisture) are implicated in the degradation. The most thorough study of the stability of doped polypyrrole was reported by Erlandsson et al. [591], who used XPS to study the reactions of polypyrrole exposed to various atmospheres at room temperature. They found little conductivity change on exposure to dry or water-saturated argon, over periods of 500 h. Exposure to dry air or oxygen produced a measureable conductivity decay, the conductivity decreasing by a factor of two over 500 h. The most significant effect occurred with water-saturated oxygen, where the conductivity decreased by a factor of 7 over the same period. XPS showed that in all cases the concentration of the BF_4^- counter-ion decreased on storage and formation of BF_3 and HF could be detected in mass-spectroscopic studies at elevated temperatures; there was little correlation between loss of BF_4^- ions and loss in conductivity. Hahn et al. [592] also found evidence for oxygen uptake into polypyrrole but were unable to detect loss of counter-ions by Auger spectroscopy. They found that the oxygen content of as-prepared films was higher on the electrode side of the films than on the outer side and suggested that some oxygen is incorporated as counter-ions during the synthesis. Wernet and Wegner [593] found that polypyrrole could be cycled between the doped and undoped states more than 1000 times in the strict absence of oxygen or water. In aqueous media, polypyrrole is susceptible to attack by OH^- at the 3-position. Oxidation can lead to $>C=O$ at the 3 position with some breakup of the conjugation.

In our own studies [568] we find that polypyrrole with BF_4^- as counter-ion loses weight quite steadily in TGA with a residue of about 30% remaining at 600 °C. The weight loss is little affected by air exposure. The conductivity of the same polymer is relatively stable but will decrease by about one order of magnitude in 600 h at 70 °C; this conductivity loss is unaffected by air exposure and is irreversible in that the polymer cannot be redoped either electrochemically or from the gas phase. In the absence of oxygen the polymer is apparently insensitive to water vapour but it loses conductivity on exposure to ammonia, the conductivity of a 2 μm thick film falling by about 50% in 40 min at room temperature; this reaction is at least partially reversible and the conductivity can be substantially restored if the ammonia is pumped out of the system. The degradation of polypyrrole is sensitive to the counter-ion; with ClO_4^- as counter-ion the conductivity decays more rapidly in air than in nitrogen at 70 °C.

Thus, although polypyrrole appears superficially to be a stable material, it apparently undergoes quite complex chemical changes on storage even at room temperature. The contrast with polyacetylene is that the conjugated backbone of the polypyrrole chain is less reactive so that there is less destruction of conjugation and less decline in conductivity.

A somewhat similar picture has emerged for polythiophene. This polymer has an ionization potential estimated to be above 5 eV, which should be high enough to protect the undoped polymer from oxygen. In practice, only the doped polymer has been studied and few detailed results exist. Tourillon and Garnier [594] claimed that the conductivity of polythiophene with $CF_3CO_3^-$ counter-ions was unchanged after 8 months storage in air and Druy et al. [562] also reported stable conductivity. In our studies [568] we find that polythiophene is more stable than polypyrrole in a TGA experiment in nitrogen. In air the two polymers have similar stability up to around 500 °C; above this temperature the polythiophene rapidly degrades away completely in air. In an accelerated test at 70 °C we find that polythiophene loses conductivity at a rather similar rate to polypyrrole, the conductivity of the BF_4^--doped material falling by about one half in 200 h, irrespective of whether the sample is exposed to air or not. The rate of decay is somewhat sensitive to the counter-ion, being more rapid for ClO_4^- and less rapid for AsF_6^- and the degradation is irreversible. In marked contrast to polypyrrole, polythiophene is sensitive to water, the conductivity falling rapidly over periods of a few hours. The rate of this process is dependent on the film thickness. The reaction of polythiophene with ammonia is very much faster than with polypyrrole but as with polypyrrole the reaction is at least partially reversible on redoping.

Takenaka et al. [595] used XPS to study deterioration of a thin polythiophene film on an ITO electrode, such as might be used in a display device. After 10^5 doping-undoping cycles at 0.5 Hz with BF_4^- counter-ions there was evidence of extensive fluorination of the polymer, probably due to decomposition of the counter-ion or its hydrolysis by traces of water in the electrolyte. Corradini et al. [596] carried out a similar study of polypyrrole, polythiophene and their analogues with ClO_4^- counter-ions and concluded that all of the polymers are unstable to cycling or to standing in contact with the electrolyte, although polypyrrole performed best.

The ionization potential of polyphenylene is around 8 eV and it is not surprising that oxidative degradation is not a problem; the undoped polymer can withstand long periods at high temperatures in air with no change in its conductivity or its ability to dope [386]. However, the high oxidation potential creates two problems. Firstly, the range of dopants with sufficient electron affinity to oxidize the polymer is limited, and there are few solvents in which the oxidation can take place without destruction of the solvent. Secondly, the doped polymer is expected to be reactive towards water and this is indeed the case [386].

9.5 Discussion

All applications of conducting polymers depend on the material being stable. As discussed above and elsewhere [568], the instability of polyacetylene to oxygen is related to the ability of oxygen to dope the polymer and of the resulting charge-transfer complex having a low activation energy to irreversible oxidation. Oxygen diffuses rapidly into the polymer which, therefore, can only be stabilized by pre-reaction with another dopant or by processing to eliminate structural imperfections. The alternative is to encapsulate the material. Other conducting polymers, such as polythiophene are much more stable to oxygen, particularly when doped, because the lower Fermi level prevents interactions with oxygen. Unfortunately, these materials are not very moisture stable in the doped state. Munstedt [597] confirmed that the whole range of common conducting polymers is unstable in accelerated tests of conductivity decay. He suggests that the carbenium ion structures, which are required to permit conduction in conjugated polymers, are incompatible with the presence of oxygen and water and that the only practical route towards conducting polymers which have environmental stabilities comparable to graphite, will be to seek structures whose band gap is intrinsically small enough to allow thermal excitation without the need for doping.

Elsenbaumer et al. [598] recently made a comparative study of stability of doped conducting polymers in air. They concluded that a combination of a stable polymer with a non-reactive dopant, such as butylthiophene-methylthiophene copolymer doped with $FeCl_3$, could give almost indefinite stability in air with an effective ceiling of 50 °C.

In summary, few, if any, of the conducting polymers have yet been demonstrated to be sufficiently stable for critical applications. Although the study of degradation processes has barely begun we can identify several processes which can contribute to instability of properties. These are: 1) Reaction of the main chain with oxygen, leading to irreversible loss of conjugation and conductivity. 2) Reaction of the main chain with its counter-ion, again leading to irreversible loss of conjugation and conductivity. 3) Reaction with oxygen or counter-ions at hetero-atoms without loss of conjugation but with modification of the electronic structure of the polymer. 4) Reversible or partly reversible undoping reactions of the conjugated system.

10 Applications of Conducting Polymers

If one asks what are the applications of conducting polymers, the short answer is "none". At the present time (July 1988), the most active field of development is in batteries. There have also been large programmes aimed at developing photovoltaic cells, chemical sensors, semiconductor devices and optical switches. A host of small groups have also investigated the feasibility of various applications. A complete survey is also very difficult because the tendency is to publish completed but unsuccessful studies.

10.1 Batteries

McDiarmid's group has put a great deal of effort into rechargeable (secondary) polyacetylene batteries [599]. Several different configurations have been tested. As set out in table 4, cells can be constructed with two polyacetylene electrodes, both p-type, both n-type, one of each, or either form against lithium metal.

The essential issues with respect to a practical secondary battery are the energy density (energy/weight ratio), the power density, spontaneous discharge rate, fractional charge recoverable (reversibility) and the lifetime. In order to compare figures for different systems it is essential to ensure that they are derived on the same basis, for

Table 4. Composition and EMF of polyacetylene cells [599, 600]

Cell	Voltage
$(CH)_x/Li$	1.9
$(CH)_x^{0.06+}/Li$	3.7
$(CH)_x^{0.024+}/(CH)_x^{0.024-}$	2.5
$(CH)_x^{0.06+}/(CH)_x^{0.06-}$	2.5
$(CH)_x^{0.05+}/(CH)_x^{0.02+}$	0.5
$(CH)_x^{0.02-}/(CH)_x^{0.02-}$	0.66

instance energy density may be given on the basis of the weight of the electrodes alone or on the basis of the whole package. The standard for comparison is the lead-acid battery which has undergone so many years of refinement that it is very hard to better.

Shirakawa polyacetylene has a very high surface area, due to the fibrillar structure, and thus should be capable of high power densities. Initial maximum power densities for $(CH)_x$/Li cells have been given [600] as 2.9 kW kg^{-1} based on the weight of the electrodes alone. Average power densities over the total discharge were 20–200 W kg^{-1} depending on the discharge rate. The slow diffusion of ions in the polyacetylene fibrils limits the discharge rate of the cell. For comparison, the power density of a lead-acid battery is about 200 W kg^{-1} on the same basis. Chiang [601] described an all-polymer battery, based on two polyacetylene electrodes and a poly(ethylene oxide) electrolyte. Running at 85 °C the power density was only 10–20 W kg^{-1}. A poly-acetylene-sodium battery with the same electrolyte gave a maximum power density of 250 W kg^{-1} and an energy density of 20 Wh kg^{-1}. The energy density of a p-type polyacetylene-lithium battery is very dependent on the extent of doping of the poly-acetylene. This is, in turn, limited by the tendency for the polymer to degrade at high dopant levels. Theoretical values are 100–200 Wh kg^{-1}, depending on discharge rate. Taking into account the expected weight of electrolyte and packing, this becomes about 80 Wh kg^{-1}, compared to 40 Wh kg^{-1} for lead-acid batteries [602, 603, 604]. An all-polymer battery has been described with polyacetylene electrodes and an electrolyte of polyvinylidenefluoride swollen with lithium perchlorate in propylene carbonate [605]. Based on the weight of the electrodes, the claimed energy density was 47 Wh kg^{-1}, the power density 3.2 kW kg^{-1}. Self-discharge was a serious problem, with 50% of the charge being lost in 5 hours. This was attributed to reactions between the dopant and polyacetylene.

The major problem with polyacetylene is its instability when doped, even in an inert atmosphere. As discussed above, degradation occurs at doping levels above 10% or with oxidising dopants such as perchlorate. This instability has been demonstrated by resonance Raman spectroscopy, during cyclic charging and discharge, which shows a loss of all long conjugation lengths [606]. Farrington has compared polyacetylene with TiS_2 as electrode materials and shown that the polyacetylene has considerably lower energy densities [607].

The Allied Chemical group have studied poly-p-phenylene batteries using pressed powder electrodes [608]. Polyphenylene against lithium gives a rather low voltage, 0.9 V dropping to 0.4 V during discharge. Hexafluoroarsenate-doped polyphenylene against lithium gives 4.4 V, but this high potential difference tends to electrolyse any solvent that is used as electrolyte. This can be resolved by use of carefully purified, liquid sulphur dioxide at −40 °C as electrolyte [609]. This cell was constructed with electrodes of electropolymerized poly-p-phenylene.

Naarman and co-workers have discussed the construction of lithium-polypyrrole batteries [610]. The initial voltage of 3.5 V droped to 2.5 V during discharge. They were able to obtain energy densities of 8–40 Wh kg^{-1}, based on total battery weight. The efficiency of charge return was initially 80%, but depended on there being some water present. During the first few cycles, this dropped to 30% as the lithium took up the water. An important factor in limiting the energy density was the need for enough electrolyte to solvate all the extra ions produced during discharge. The self-discharge rates were 1% per day, comparable to other rechargeable batteries. The higher repeat-unit molecular weight of polymers other than polyacetylene reduces the possible energy density.

A polypyrrole-electrolyte-polypyrrole battery has been described [611] but the n-doped polypyrrole is unstable. This can be avoided by using a polypyrrole-polyanion anode, where the charge and discharge depend upon small cations moving into and out of the electrode [612]. A polythiophene-polythiophene battery has also been described [135]. Polyaniline has been studied as a potential battery cathode, for use with an aqueous electrolyte [613, 614].

Thus, it seems that the limits have been defined on conducting polymer secondary batteries and that they are potentially competitive with existing batteries but not vastly superior. A recent review by Shacklette et al. [615] discusses a variety of polymers as potential electrodes, and the question of electrolyte stability, and ends on a reasonably optimistic note.

10.2 Semiconducting Devices and Photovoltaics

Kanicki [616] has recently published an extensive review of the semiconducting properties of polyacetylene. We will summarize the field only briefly here.

Schottky diodes can be prepared by forming a sandwich of polyacetylene between two dissimilar metals. One metal, such as gold, must form an ohmic contact while the other interface is to a low work function metal, for instance to lead or indium, and so is rectifying. Studies of the behaviour of such diodes give a lot of information about the electronic properties of polyacetylene. Polyacetylene Schottky diodes have a rectification ratio in the range of 100 to 500. If one electrode is made transparent, the diode will act as a photovoltaic cell when illuminated. In white light of roughly solar intensity, the cells gave open circuit voltages from 80–700 mV and power conversion efficiencies in the range of $10^{-4}\%$ to 0.2%. These figures compare with efficiencies of 20% and 5% for silicon and amorphous silicon cells respectively. Much of the poor efficiency arises from a low quantum yield. The charge carriers generated by absorption of light have a low mobility (about 3×10^{-6} cm^2 V^{-1} s^{-1}) and a low lifetime (about 5×10^{-5} s) and thus tend to recombine before they can diffuse across the junction. Doping the polymer increases the conductivity but reduces the photovoltaic efficiency. A polythiophene Schottky junction gave a maximum efficiency of $3 \times 10^{-3}\%$ [617].

In principle diodes could be made by joining, or forming in-situ, a p-type/n-type polymer junction. In polyacetylene such junctions are unstable due to counter-ion migration and reaction. A p–n junction which is claimed to be stable for a week in air, has been made by ion implantation of sodium into lightly iodine-doped polyacetylene [618]. In general ion implantation seems to graphitize conducting polymers. The rectification ratio was only 12. A p–n junction has also been made by polymerization of n-type polythiophene onto p-type polypyrrole, the rectification ratio was 15 [619]. Junctions of n-type polyacetylene with polypyrrole or polythiophene have been prepared. A rectification ratio of 150 is claimed, with photovoltaic efficiencies of $10^{-4}\%$ [620]. There has been an extensive study of heterojunctions of polyacetylene on n-type cadmium sulphide [621]. The diodes formed good rectifiers but the photovoltaic efficiency is $3 \times 10^{-3}\%$. Other heterojunctions were more successful, polypyrrole/n-type silicon reached 1.2% and lightly AsF_5^--doped polyacetylene/n-type silicon reached 4.3%, but much of the activity may be within the silicon rather than in the polymer.

A polyacetylene field-effect transistor has been described [622] but the response time is slow, apparently because the carrier mobility is low. An FET has been made from polythiophene but source-drain currents were less than 20 nA for drain voltages up to 50 V. The hole mobility was very low, calculated to be 2×10^{-5} cm^2 V^{-1} s^{-1} [623].

The attraction of semiconducting polymers is the potential for printing large sheets of cheap photovoltaic cells or transistors. This is only viable if the properties are comparable with those of silicon, and the possible advantage has been partly removed by the development of good amorphous silicon films. The two major stumbling blocks to exploitation are the instability of polyacetylene and the low carrier mobilities. Undoped polyacetylene is not stable enough to be used in air and thus any device must be completely encapsulated. A thick polymer shell or a glass or metal box would be necessary to exclude oxygen for 5 years. This air sensitivity also makes preparation of the devices inconvenient. Other polymers are more stable but do not have the intrinsic conductivity needed to make devices.

Recent studies, described above [16], have shown much increased conductivity in highly oriented polyacetylene. This implies that the carrier mobility has increased because of the greater extent on intra-chain conduction due to removal of kinks during drawing. If this proves to be the case, it may well be that big improvements in photovoltaic cells and devices can be achieved but the drawing process does not really lend itself to device manufacture, so other methods of attaining similar properties, such as epitaxial film formation, would have to be found. In addition, the fibrillar morphology is not well suited to producing good junctions.

Yoshino et al. [624, 625] and Gazard [626] have decribed electrochromic devices which exploit the colour change in polypyrrole or polythiophene films on doping. These films can switch in 0.2 s at a voltage of ± 1.4 V, and will operate for 5×10^5 cycles, comparable with current tungstic oxide devices. Faster switching occurs at higher potentials but at the cost of more rapid degradation. The switching time is limited by the diffusion rate in the film.

10.3 Catalysis

There is a strong interest in the use of modified conducting polymers as catalytic electrodes. Noufi [627] and Bull et al. [628] have included RuO_2 and phthalocyanines in polypyrrole to catalyse photooxidation of water and the reduction of O_2 to H_2O_2, respectively. Incorporation of a sulphonated cobalt porphyrin as an anion in polypyrrole is claimed to give good activity for oxygen reduction [629]. Polythiophene with incorporated silver and platinum particles has been used to catalyse the reduction of H^+ [630]. The silver, at 80 wt% of the polymer, was aggregated on the polymer fibres and enhanced the conductivity, while the Pt was located near the film surface and was responsible for the catalytic activity. The activity was 50% higher than for a Au-Ag-Pt electrode. Platinum and other active particles have been incorporated into polypyrrole composites by dispersing the particles into polyvinylchloride and then electropolymerising the pyrrole into the PVC [631].

10.4 Sensors

In view of their sensitivity to oxidation and to doping gases, it would seem natural to use conducting polymers as chemical sensors. Polyacetylene, for instance, is readily doped by oxygen which induces a large increase in conductivity. At present, many gas sensors are ceramics where conductivity changes occur, often in rather poorly characterized surface layers. High temperatures are needed for their operation. Degradation of conducting polymers is again a major drawback, and slow diffusional processes limit response times, but the field is still very much in its infancy. A polypyrrole sensor for NO_2 and NH_3 has been made by electropolymerization of pyrrole onto a pattern of interdigitated electrodes [632]. On exposure to 0.1% NH_3, the conductivity changed by 2.5%, but the response time was greater than 15 minutes. It would thus be necessary to sense a rate of change of conductivity, or to develop a more open morphology which could respond faster.

The initial hope for conducting polymers was that they would replace metals in applications as simple conductors. The interest for conducting coatings and radio-frequency shielding remains, but is so far held up by problems of instability and cost. Much work in this area is not published because it is carried out by the defence industries.

11 Summary

The initial promise of conducting polymers was a conductivity like copper combined with a processibility like polyethylene. Work over the last 10 years has gone a long way toward this goal, but there is still a long way to go.

A vital part of the early development of plastics was the evolution of methods of polymerization to reliable high molecular weight products. Structure and mechanical properties are very sensitive to molecular weight and to small fractions of branches or cross-links. Polyacetylene is made by a method which is analogous to the commercial polyolefins and has been shown to be a high molecular weight linear polymer. Most of the other systems which have been studied, including polythiophenes, polypyrroles and polyphenylene, have structures which are uncertain as to both molecular weight and defects. Coupling reactions are particularly ineffective for producing high molecular weight products and the many aromatic polymers produced by these methods are often little more than oligomeric organic glasses. The refinement needed to produce good polymers is very demanding of time and effort, so it is unlikely that high-grade materials will appear until some applications have developed. The conducting properties have not, so far, proved to be very sensitive to the molecular weight, but many of the properties of importance during the use of these polymers will be very dependent of the chain microstructure.

The intractability of the conducting polymers makes characterization difficult and this in turn slows the development of better polymers. The precursor routes are very attractive because they provide intermediate polymers which can be properly characterized. A precursor for polypyrrole or polythiophene would greatly enhance our ability to understand the structure of the polymers produced electrochemically.

Polymer crystal structure is not very dependent on molecular weight or chain microstructure. However, the ratio of crystal to amorphous content is very sensitive to chain regularity. A few per cent of copolymer units or of interruptions to tacticity can eliminate the crystallinity of a vinyl polymer. The establishment of the two-phase model for the structure of crystalline polymers was a long and hard-fought battle through the 1940's and 1950's. While this is well established for flexible chain polymers, it is not clear that it also applies to stiff chains like the conducting polymers. In the era of plastics development, there were few stiff chain materials and their structure could be ignored. Now we have to go back and give careful thought to what kind of phase behaviour would be expected for a long semi-flexible chain. Paracrystalline models were discarded by most workers, as descriptions for the vinyl polymers, but may be applicable to the conducting polymers.

The only polymer which has been subjected to systematic studies of mechanical properties is polypyrrole, and even here there are few data. Increasing the polymer molecular weight would be expected to lead to improved toughness, particularly in the very short-chain materials such as Kovacic polyphenylene.

The production of highly conducting, oriented polyacetylenes has demonstrated that most normal measurements of conductivity are dominated by chain-chain hopping of the charge carriers. The observed order of conductivity in doped polymers (Poly-acetylene = polyphenylene > polypyrrole > polythiophene) must reflect chain pack-ing effects. We thus have no idea how the conductivities of other, hypothetical, highly oriented polymers would compare. The production of high levels of orientation and perfection in other polymers must be a high research priority, as must be the elucidation of the origin of the special properties of the highly oriented polyacetylenes produced in silicone oil or liquid crystal solvents. If such materials become available, we would also expect much more sensitivity of conductivity to molecular weight and regularity than is now observed.

Of the large number of possible applications for conducting polymers, the most promising for the first significant exploitation is secondary batteries. The systems being developed are somewhat better than the lead-acid battery in terms of pure performance, though not as good as several other promising future systems. Whether polymer batteries do come into use will depend on detailed requirements of particular applications.

Polymer science is underdeveloped in terms of descriptions of the structure and properties of stiff-chain polymers. The conducting polymers fall mostly within this blind spot. They also present a number of novel possibilities such as the conversion from a flexible-chain precursor to a rigid-chain polymer, and the conversion between doped and undoped states in the soluble polythiophenes. Likewise, solid-state physics has yet really to tackle the transport of electrons in, and between, disordered, twisted chains. For each of the disciplines involved, the explosion of interest in conducting polymers has brouht a host of new question and new ideas. The process is far from over.

12 References

1. Berets DJ, Smith DS (1968) Trans. Faraday Soc. *64*: 823
2. Shirakawa H, Ikeda S (1971) Polym. J. *2*: 231
3. Shirakawa H, Ikeda S (1974) J. Polymer Sci., Chem. *12*: 929

4. Shirakawa H, Louis EJ, MacDiarmid AG, Chiang CK, Heeger AJ (1977) J. Chem. Soc.: Chem. Commun: 578
5. Wegner G (1986) Makromol. Chem.: Makromol. Symp. *1*: 151
6. Chien JCW (1984) Polyacetylene: Chemistry, physics and materials science, Academic, New York
7. Natta G, Mazzanti G, Corradini P (1958) Atti Acad. Naz. Lincei Rend. Sci. Fis. Mat. Nat. *25*: 2
8. Ito T, Shirakawa H, Ikeda S (1974) J. Polymer Sci., Chem. *12*: 11
9. Clarke TC, Yannoni CS, Katz TJ (1983) J. Amer. Chem. Soc. *105*: 7787
10. Clarke TC, Yannoni CS (1983) Phys. Rev. Lett. *51*: 1191
11. Saxman AM, Liepins R, Aldissi M (1985) Prog. Polymer Sci. *11*: 57
12. Cao Y, Quian R, Wang F, Zhao X (1982) Makromol. Chem.: Rapid Commun. *3*: 687
13. Catellani M, Destri S, Bolognesi A (1986) Makromol. Chem. *187*: 1354
14. Kminek I, Trekoval J (1986) Makromol. Chem.: Rapid Commun. *7*: 53
15. Theophilou N, Aznar R, Munardi A, Sledz J, Schue F, Naarmann H (1987) J. Macromol. Sci. *A24*: 797
16. Munardi A, Theophilou N, Aznar R, Sledz J, Schue F, Naarmann H (1987) Makromol. Chem. *188*: 395
17. Naarmann H, Theophilou N (1987) Synth. Met. *22*: 1
18. Basescu N, Liu ZX, Moses D, Heeger AJ, Naarmann H, Theophilou N (1987) Nature *327*: 403
19. Ivin KJ (1983) Olefin metathesis, Academic, New York; Dragutan V, Balaban AT, Dinonie M (1985) Olefin metathesis and ring-opening polymerization of cycloolefins, Wiley, New York
20. Amass A, Beevers MS, Farren TR, Stowell JA (1987) Makromol. Chem.: Rapid Commun. *8*: 119
21. Theophilou N, Munardi A, Aznar R, Sledz J, Schue F, Naarmann H (1987) Eur. Polymer J. *23*: 15
22. Luttinger LB (1960) Chem. Ind. Lond., 1135
23. Luttinger LB (1962) J. Org. Chem. *27*: 1591
24. Lieser G, Wegner G, Muller G, Enkelmann V, Meyer WH (1980) Makromol. Chem.: Rapid Commun. *1*: 621, 627
25. Korshak YV, Korshak VV, Kanischka G, Hocker H (1985) Makromol. Chem.: Rapid Commun. *6*: 685
26. Klavetter FL, Grubbs RH (1987) Polymer Prepr. *27(2)*: 425
27. Heaviside J, Hendra PJ, Tsai P, Cooney RP (1978) J. Chem. Soc.: Faraday Trans. I, *74*: 2542
28. Murakami S, Tabata M, Sohma J, Hatano M (1984) J. Appl. Polymer Sci. *29*: 3445
29. Feast WJ (1986) In: Skotheim TA (ed) Handbook of conducting polymers, Dekker, New York, Vol 1, Ch. 1
30. Gibson HW, Pochan JM (1984) Encyclopaedia of polymer science and engineering, Wiley, New York, 2nd. Ed., *1*: 87
31. Simionescu CI, Percec V (1982) Prog. Polymer Sci. *8*: 133
32. Hankin AB, North AM (1967) Trans. Far. Soc. *63*: 1525
33. Ehrlich P, Anderson W (1986) In: Skotheim TA (ed) Handbook of conducting polymers, Dekker, New York, Vol. 1, Ch. 12
34. Cukor P, Krugler JI, Rubner MF (1981) Makromol. Chem. *182*: 165
35. Wentworth SE, Bergquist PR (1985) J. Polymer Sci.: Chem. *23*: 2197
36. Chien JCW, Wnek GE, Karasz FE, Hirsch JA (1981) Macromolecules *14*: 479
37. Zeigler JW (1985) Polymer Prepr. *25(2)*: 223
38. Chien JCW, Carlini C (1984) Makromol. Chem.: Rapid Commun *5*: 597; (1985) J. Polymer Sci.: Chem. *23*: 1383
39. Leclerc M, Prud'homme RE (1987) Polymer Bull. *18*: 159
40. Bredas JL, Street GB, Themans B, Andre JM (1985) J. Chem. Phys. *83*: 1323
41. MacDiarmid A, Heeger A (1980) J. Org. Coat. Plast. Chem. *43*: 853
42. Gibson HW (1985) Polymer *25*: 3
43. Butler GB, Corfield GB, Aso C (1975) Prog. Polymer Sci. *4*: 71
44. Gibson, HW, Bailey FC, Epstein AJ, Rommelmann H, Kaplan S, Harbour J, Yang X, Tanner DB, Yang XQ, Pochan JM (1983) J. Amer. Chem. Soc. *105*: 4417

45. Gibson HW (1984) New monomers and polymers, Plenum, New York, p. 381
46. Gibson HW, Weagley RJ (1986) Brit. Polymer J. *18*: 120
47. Speight JG, Kovacik P, Koch FW (1971) J. Macromol. Sci.: Revs. *C5*: 295
48. Kovacik P, Jones MB (1987) Chem. Rev. *87*: 357
49. Noren GO, Stille JK (1971) Makromol. Revs. *5*: 385
50. Edwards G, Goldfinger G (1955) J. Polymer Sci. *16*: 589
51. Berlin AA, Liogonku VI, Parnini VP (1961) J. Polymer Sci. *55*: 675
52. Yamamoto T, Hayashi Y, Yamamoto A (1978) Bull. Chem. Soc. Japan *51*: 2091
53. Taylor SK, Bennett SG, Khoury I, Kovacik P (1981) J. Polymer Sci.: Letters *19*: 85
54. Balaban AT, Nenizescu CD (1965) In: Olah GA (Ed) Friedel-Crafts and related reactions, Interscience, New York, vol 2
55. Kovacik P, Kyriakis A (1962) Tetrahedron Letters, 467; (1963) J. Amer. Chem. Soc. *85*: 454
56. Tiecke B, Bubeck C, Lieser G (1982) Makromol. Chem.: Rapid Commun. *3*: 261
57. Kitajima N, Hakone Y, Ono Y (1982) Chem. Lett., 871
58. Kovacik P, Wu C (1960) J. Polymer Sci. *47*: 45
59. Kovacik P, Lange RM (1963) J. Org. Chem. *28*: 968
60. Teraoka F (1980) J. Macromol. Sci.: Phys. *B18*: 73
61. Shacklette LW, Eckhardt EH, Chance RR, Miller GG, Ivory DM, Baughmann RH (1980) J. Chem. Phys. *73*: 4098
62. Kossmehl G, Chatzitheodourou G (1981) Makromol. Chem.: Rapid Commun. *2*: 551
63. Schopov I, Kossmehl G (1987) Polymer Commun. *28*: 34
64. Aldissi M, Liepins R (1984) J. Chem. Soc.: Chem. Commun 255
65. Engstrom GG, Kovacik P (1977) J. Polymer Sci. Phys. *15*: 2453
66. Mano EB, Alves L (1972) J. Polymer Sci.: Chem. *10*: 655
67. Hsing CF, Khoury I, Bezoari MD, Kovacik P (1982) J. Polymer Sci.: Chem. *20*: 3313
68. Hill HW Jr, Brady DC (1982) Kirk-Othmer encyclopaedia of chemical technology, Wiley-Interscience, New York, p 793
69. Clarke TC, Kanazawa KK, Lee VY, Rabolt JF, Reynolds JR, Street GB (1982) J. Polymer Sci.: Physics *20*: 117
70. Frommer JE, Elsenbaumer RL, Eckhardt H, Chance RR (1983) J. Polymer Sci.: Lett. *21*: 39
71. Cleary JW (1986) Polymer Sci. Technol. *31*: 173
72. Rajan CR, Ponrathnam S, Nadkarni VM (1986) J. Appl. Polymer Sci. *32*: 4479
73. Lenz RW, Handlovits CE, Smiths HA (1962) J. Polymer Sci. *58*: 351
74. Shacklette LW, Elsenbaumer RL, Chance RR, Eckhardt H, Frommer JE, Baughman RH (1981) J. Chem. Phys. *75*: 1919
75. Friend RH, Giles RM (1984) J. Chem. Soc.: Chem. Commun.: 1101
76. Tsukamoto J, Fukuda S, Tanaka K, Yamabe T (1987) Synth. Met. *17*: 673
77. Chance RR, Shacklette LW, Miller GG, Ivory DM, Sowa JM, Baughmann RH (1980) J. Chem. Soc.: Chem. Commun.: 347
78. Jen KY, Lakshmikantham MV, Albeck M, Cava MP, Huang WS, MacDiarmid AG (1983) J. Polymer Sci.: Lett. *21*: 441
79. Sandman DJ, Rubner M, Samuelson L (1982) J. Chem. Soc.: Chem. Commun.: 1133
80. Salmon M, Kanazawa KK, Diaz AF, Krounbi M (1982) J. Polymer Sci.: Lett. *20*: 187
81. Armes SP (1987) Synth. Met. *20*: 365
82. Walker JA, Warren LF, Witucki EF (1987) Polymer Prepr. *28(2)*: 256
83. Walker JA, Warren LF, Witucki EF (1988) J. Polymer Sci. Chem. *26*: 1285
84. Mermilliod N, Tanguy J, Petiot F (1986) J. Electrochem. Soc. *133*: 1073
85. Neoh KG, Tan TC, Kang ET (1988) Polymer *29*: 553
86. Kang ET, Neoh KG, Tan TC, Ong YK (1987) J. Macromol. Sci.: Chem. *A24*: 631
87. Kang ET, Neoh KG, Tan TC, Ong YK (1987) J. Polymer Sci.: Chem. *25*: 2143
88. Chao TH, March J (1988) J. Polymer Sci.: Chem. *26*: 743
89. Bocchi V, Gardini GP (1986) J. Chem. Soc.: Chem. Commun.: 148
90. Bjorklund R, Lundstrom I (1984) J. Electron. Mater. *13*: 211
91. Ojio T, Miyata S (1986) Polymer J. *18*: 95
92. Mohammadi A, Hassan MA, Liedberg B, Lundstrom I, Salanek WR (1986) Synth. Met. *14*: 189

93. Mohammadi A, Lundstrom I, Salanek WR, Inganas O (1986) Chemtronics *1*: 171
94. Yamamoto T, Senichica K, Yamamoto A (1980) J. Polymer Sci.: Lett. *18*: 9
95. Kossmehl C, Chatzihedorou G (1981) Makromol. Chem.: Rapid Commun. *2*: 551
96. Kobayashi M, Chen J, Chung TC, Moraes F, Heeger AJ, Wudl F (1984) Synth. Met. *9*: 77
97. Cunningham DD, Laugren-Davidson L, Mark HB, Pham CV, Zimmer H (1987) J. Chem. Soc.: Chem. Commun.: 1027
98. Amer A, Zimmer H, Mulligan KJ, Mark HB, Pons S, McAleer JF (1984) J. Polymer Sci.: Lett. *22*: 77
99. Pham CV, Czerwinski A, Zimmer H, Mark HB (1986) J. Polymer Sci., Lett. *24*: 103
100. Berlin A, Pagani GA, Sannicolo F (1986) J. Chem. Soc.: Chem. Commun.: 1663
101. Sugimoto R, Takeda S, Gu HB, Yoshino K (1986) Chem. Exp. *1*: 635
102. Wnek GE, Chien JCW, Karasz FE, Lillya CP (1979) Polymer *20*: 1441
103. Kossmehl G (1986) In: Skotheim A (Ed) Handbook of conducting polymers, Dekker, New York, vol 1 Ch 10
104. Kossmehl G, Beimling P, Manecke G (1982) Makromol. Chem. *183*: 2771
105. Droske JP, Johnson PM, Engen PT (1987) Macromolecules *20*: 461
106. Kossmehl G, Yaridjanian A (1981) Makromol. Chem. *182*: 3419
107. Yamazaki N (1969) Adv. Polymer Sci. *6*: 377
108. Badani S, Parravano G (1983) In: Baizer MM, Lund H (Eds) Organic Electrochemistry, Dekker, New York
109. Naarman H (1983) US Patent, 4,468,291
110. Kornicker WA (1969) US Patent, 3,474,012
111. Farafnov V, Grovu M, Simionescu CI (1977) J. Polymer Sci.: Chem. *15*: 2041
112. Subramanian RV (1979) Adv. Polym. Sci. *33*: 43
113. Chen SA, Shy HJ (1985) J. Polymer Sci.: Chem. *23*: 2441
114. Chandler GK, Pletcher D (1985) Spec. Period. Report Electrochem. *10*: 117
115. Dall'Olio A, Dascola Y, Varacca V, Bocchi V (1968) Compt. Rend. *C267*: 433
116. a) Diaz AF, Kanazawa KK, Gardini GP (1979) J. Chem. Soc.: Chem. Commun.: 635; b) Kanazawa KK, Diaz AF, Geiss RH, Gill WD, Kwak JF, Rabolt JF, Street GB (1979) J. Chem. Soc.: Chem. Commun. 854; c) Kanazawa KK, Diaz AF, Gill WD, Grant PM, Street GB, Gardini GP, Kwak JF (1980) Synth. Met. *1*: 329; d) Diaz AF, Castillo JI (1980) J. Chem. Soc.: Chem. Commun.: 397
117. Diaz AF (1981) Chem. Scripta *17*: 145
118. Clarke TC, Clarke JC, Street GB (1983) IBM J. Res. Dev. *27*: 313
119. Street GB, Clarke TC, Krounbi M, Kanazawa K, Lee V, Pfluger P, Scott J, Weiser G (1982) Mol. Cryst. Liq. Cryst *83*, 253
120. Lindenberger H, Schafer-Siebert D, Roth S, Hanack M (1987) Synth. Met. *18*: 37
121. Mengoli G, Musiani M, Fleischmann M, Pletcher D (1984) J. Appl. Electrochem. *14*: 285
122. Warren LF, Anderson DP (1987) J. Electrochem. Soc. *134*: 101
123. Qian R, Qiu J (1987) Polymer J. *19*: 157
124. Qian R, Qiu J, Shen D (1987) Synth. Met. *18*: 13
125. Wernet W, Monkenbusch M, Wegner G (1984) Makromol. Chem.: Rapid Commun. *5*: 157
126. Wernet W, Monkenbusch M, Wegner G (1985) Mol. Cryst. Liq. Cryst. *18*: 193
127. Shimidzu T, Ohtani A, Iyoda T, Honda K (1986) J. Chem. Soc.: Chem. Commun.: 1415
128. Glatzhofer DT, Ulanski J, Wegner G (1987) Polymer *28*: 449
129. Okano M, Itoh K, Fujishima A, Honda K (1987) J. Electrochem. Soc. *134*: 837
130. Diaz AF, Castillo J, Kanazawa KK, Salmon M (1982) J. Electroanal. Chem. *133*: 233
131. Street GB, Clarke TC, Geiss RH, Lee YV, Nazzal A, Pfluger P, Scott JC (1983) J. Phys.: Colloq. *C3*: 599
132. Pfluger P, Street GB (1983) J. Phys.: Colloq. *C3*: 609
133. Tourillon G, Garnier F (1982) J. Electroanal. Chem. *135*: 173; (1983) J. Phys. Chem. *87*: 2289; (1984) J. Electroanal. Chem. *161*: 51, 407
134. Kaneto K, Kohno Y, Yoshino K, Inuishi Y (1983) J. Chem. Soc.: Chem. Commun.: 382
135. Kaneto K, Yoshino K, Inuishi Y (1982) Japan J. Appl. Phys. *L21*: 567; (1983) *L22*: 412
136. Tourillon G (1986) In: Skotheim TA (Ed) Handbook of conducting polymers, Dekker, New York, Vol. 1, Ch. 9
137. Garnier F, Tourillon G, Gazard M, Dubois JC (1983) J. Electroanal. Chem. *148*: 299

138. Hotta S, Hosaka T, Shimotsuma W (1983) Synth. Met. 6: 317
139. Sato M, Tanaka S, Kaeriyama K (1985) J. Chem. Soc.: Chem. Commun.: 713
140. Roncali J, Lemaire M, Garreau R, Garnier F (1987) Synth. Met. 18: 139
141. Czerwinski A, Zimmer H, Pham CV, Mark HB Jr (1985) J. Electrochem. Soc. 132: 2669
142. Osawa S, Ito M, Tanaka K, Kuwano J (1987) Synth. Met. 18: 145
143. Waltman RJ, Bargon J, Diaz AF (1983) J. Phys. Chem. 87: 1459
144. Audebert P, Bidan G (1985) J. Electroanal Chem. 190: 129
145. Roncali J, Garnier F, Lemaire M, Garreau R (1986) Synth. Met. 15: 323
146. Funt BL, Lowen SV (1985) Synth. Met. 14: 129
147. Krische B, Hellberg J, Lilja C (1987) J. Chem. Soc.: Chem. Commun.: 1477
148. Kaneto K, Agawa H, Yoshino K (1987) J. Appl. Phys. 61: 1197
149. Tanaka S, Sato M, Kaeriyama K (1985) Polymer Commun. 26: 303
150. Kaeriyama K, Sato M, Tanaka S (1987) Synth. Met. 18: 233
151. Sato M, Tanaka S, Kaeriyama K (1987) J. Chem. Soc.: Chem. Commun.: 1725
152. Sato M, Tanaka S, Kaeriyama K, Tomonaga F (1987) Polymer 28: 1071
153. Lemaire M, Delabouglise D, Garreau R, Guy A, Roncali J (1988) J. Chem. Soc.: Chem. Commun.: 659
154. Fauvarque JF, Petit MA, Pfluger F, Jutand A, Chevrot C (1983) Makromol Chem.: Rapid Commun. 4: 455
155. Favarque JF, Digua A, Petit MA, Savard J (1985) Makromol. Chem. 186: 2415
156. Favarque JF, Petit MA, Digua A (1987) Makromol. Chem. 188: 1833
157. Osa T, Yildiz A, Kuwana T (1969) J. Amer. Chem. Soc. 41: 3994
158. Delamar M, Lacaze PC, Dumousseau JY, Dubois JE (1982) Electrochim. Acta 27: 61
159. Brilmeyer G, Jasinski R (1982) J. Electrochem. Soc. 129: 1950
160. Dietrich M, Mortensen J, Heinze J (1986) J. Chem. Soc.: Chem. Commun: 1131
161. Shepard AF, Dannels BF (1966) J. Polymer Sci. A1,4: 511
162. Rubinstein I (1983) J. Electrochem. Soc. 130: 1506; (1983) J. Polymer Sci.: Chem. 21: 3035
163. Kaeriyama K, Sato M, Someno K, Tanaka S (1984) J. Chem. Soc.: Chem. Commun.: 1199
164. Satoh M, Kaneto K, Yoshino K (1985) J. Chem. Soc.: Chem. Commun.: 1629
165. Satoh M, Tabata M, Kaneto K, Yoshino K (1985) Polymer Commun. 26: 356
166. Ohsawa T, Nishihara H, Aramaki K, Takeda S, Yoshino K (1987) Polymer Commun. 28: 140
167. Ohsawa T, Inoue T, Takeda S, Kaneto K, Yoshino K (1986) Polymer Commun. 27: 247
168. Ohsawa T, Yoshino K (1987) Synth. Met. 17: 601
169. Satoh M, Uesugi F, Tabata M, Kaneto K, Yoshino K (1986) J. Chem. Soc.: Chem. Commun.: 551
170. Satoh M, Tabata M, Uesugi F, Kaneto K, Yoshino K (1987) Synth. Met. 17: 595
171. Satoh M, Uesugi F, Tabata M, Kaneto K, Yoshino K (1986) J. Chem. Soc. Chem. Commun.: 979
172. Waltman RJ, Diaz AF, Bargon J (1985) J. Electrochem. Soc. 132: 631
173. Mohilner DM, Adams RN, Argersinger WJ (1962) J. Amer. Chem. Soc. 84: 3618
174. de Surville R, Josefowicz M, Yu LT, Perichon J, Buvet R (1968) Electrochim. Acta 13: 1451; (1969) Comp. Rend., Ser. C.: 1346
175. MacDiarmid AG, Chiang JC, Halpern M, Huang WS, Mu SL, Somasiri NLD, Wu W, Yaniger SI (1985) Mol. Cryst. Liq. Cryst. 121: 173
176. Travers JP, Chroboczek J, Devreux F, Genoud F, Nechtschein M, Sayed AM, Genies EM, Tsintavis C (1985) Mol. Cryst. Liq. Cryst. 121: 195
177. Genies EM, Sayed AM, Tsintavis C (1985) Mol. Cryst. Liq. Cryst. 121: 181
178. Ohsaka T, Ohnuki Y, Oyama N, Katagiri G, Kamisako K (1984) J. Electroanal. Chem. Interfacial Electrochem. 161: 399
179. MacDiarmid AG, Chiang JC, Richter AF, Epstein AJ (1987) Synth. Met. 18: 285
180. Wnek GE (1986) Polymer Prepr. 27(1): 277
181. Hjertberg T, Salanek W, Lundstrom I, Somasiri NLD, MacDiarmid AG (1985) J. Polymer Sci.: Lett. 23: 503
182. Vachon D, Angus RO, Lu FL, Nowak M, Liu ZX, Schaffer H, Wudl F, Heeger AJ (1987) Synth. Met. 18: 297
183. Kitani A, Kaya M, Yano J, Yoshikawa K, Sasaki K (1987) Synth. Met. 18: 341

184. Chiang JC, MacDiarmid AG (1986) Synth. Met. *13*: 193
185. Huang WS, MacDiarmid AG, Epstein AJ (1987) J. Chem. Soc.: Chem. Commun: 1785
186. McManus PM, Cushman RJ, Yang SC (1987) J. Phys. Chem. *91*: 744
187. McManus PM, Cushman RJ, Yang SC (1987) Makromol Chem.: Rapid Commun. *8*: 69
188. Hancock LF, Hilker B, Chapman W, Gordon B (1986) Polymer Prepr. *27(1)*: 359
189. Gordon B, Hancock LF (1987) Polymer *28*: 585
190. Diaz AF, Castillo JI, Logan JA, Lee WY (1981) J. Electroanal. Chem. *129*: 115
191. Oyama N, Ohsaka T, Hirokawa T, Suzuki T (1987) J. Chem. Soc.: Chem. Commun.: 1133
192. Ohsaka T, Yoshimura F, Hirokawa T (1987) J. Polymer Sci.: Lett. *25*: 395
193. Kumar N, Malhotra BD, Chandra S (1985) J. Polymer Sci.: Lett. *23*: 57
194. Kanawa KK, Diaz AF, Krounbi MT, Street GB (1981) Synth. Met. *4*: 119
195. Reynolds JR, Poropatic PA, Toyooka RL (1987) Macromolecules *20*: 958
196. Reynolds JR, Poropatic PA, Toyooka RL (1987) Synth. Met. *18*: 95
197. Inganas O, Liedberg B, Chang-Ru W (1985) Synth. Met. *11*: 239
198. Sundaresan NS, Basak S, Pomerantz M, Reynolds JR (1987) J. Chem. Soc.: Chem. Commun: 621
199. Naitoh S, Sanui K, Ogata N (1986) J. Chem. Soc.: Chem. Commun. 1348
200. Mcleod GG, Mahboubian-Jones MGB, Pethrick RA, Watson SD (1986) Polymer *27*: 455
201. Ferrarsi JP, Skiles GD (1987) Polymer *28*: 179
202. Danielli R, Ostoja P, Tiecco M, Zambioni R, Taliani C (1986) J. Chem. Soc.: Chem. Commun.: 1473
203. Mitsuhara T, Kaeriyama K, Tanaka S (1987) J. Chem. Soc.: Chem. Commun.: 765
204. Tanaka S, Sato M, Kaeriyama K (1987) J. Macromol. Sci. *A24*: 749
205. Berlin A, Wernet W, Wegner G (1987) Makromol. Chem. *188*: 2963
206. Bragadin M, Cescon P, Berlin A, Sannicolo F (1987) Makromol. Chem. *188*: 1425
207. Tanaka S, Sato M, Kaeriyama K (1985) Makromol. Chem. *186*: 1685
208. Jen K-Y, Eckhardt H, Tow TR, Shacklette LW, Elsenbaumer RL (1988) J. Chem. Soc.: Chem. Commun.: 215
209. Becher M, Mark HF Jr (1961) Angew. Chem. *73*: 641
210. Mainthia SB, Kronick PL, Labes MM (1962) J. Chem. Phys. *37*: 2509
211. Roth H, Heise L (1954) Kunststoffe *44*: 8, 341; (1957) Fuchs N, Makromol. Chem. *22*: 1; (1958) Bohrer J, Trans. N.Y. Acad. Sci. *2*: 20
212. Gerrard DL, Maddams WF (1977) Macromolecules *10*: 1221
213. Tsuchida E, Shih C-N, Shinohara I, Kambara S (1964) J. Polymer Sci. *A2*: 3347
214. Soga K, Nakamaru M, Kobayashi Y, Ikeda S (1983) Synth. Met. *6*: 275
215. Kise H, Sugihara M, He F-F (1985) J. Appl. Polymer Sci. *30*: 1133
216. Howang KT, Iwamoto K, Seno M (1986) Makromol. Chem. *187*: 611
217. Kise H (1982) J. Polymer Sci.: Chem. *20*: 3189
218. Decker C (1987) J. Polymer Sci.: Lett. *25*: 5
219. Yie-Shun C, Jagur-Grodzinski J, Vofsi D (1985) J. Polymer Sci. Chem. *23*: 1193
220. Kise H, Ogata H (1980) J. Polymer Sci. Chem. *21*: 3443
221. Faingold-Murshak V, Karvaly B (1985) Polymer Commun. *26*: 358
222. Diaz F, McCarthy TJ (1985) J. Polymer Sci.: Chem. *23*: 1057
223. Iqbal Z, Ivory DM, Szobota JS, Elsenbaumer RL, Baughman RH (1986) Macromolecules *19*: 2992
224. Miyata M, Okanishi K, Takemoto K (1986) Polymer J. *18*: 185
225. Tsutsumi H, Okanishi K, Miyata M, Takemoto K (1987) Makromol. Chem.: Rapid Commun. *8*: 135
226. Schomaker JA, Tolbert LM (1988) Polymer Prepr. *29(1)*: 268
227. Edwards JH, Feast WJ (1980) Polymer *21*: 595
228. Wilhelm D, Strohmaier D, Kleeberg W (1986) Angew. Makromol. Chem. *141*: 141
229. Foot PJS, Calvert PD, Billingham NC, Brown CS, Walker NS, James DI (1986) Polymer *27*: 448
230. Brown CS, Vickers ME, Foot PJS, Calvert PD, Billingham NC (1986) Polymer *27*: 1719
231. Bott DC, Chai CK, Walker NS, Feast WJ, Foot PJS, Calvert PD, Billingham NC, Friend RH (1986) Synth. Met. *14*: 245
232. Feast WJ, Winter NJ (1985) J. Chem. Soc.: Chem. Commun.: 202

233. Wnek GE, Chien JCW, Karasz FE, Lillya CP (1979) Polymer 20: 1141
234. Horhold H-H, Helbig M (1987) Makromol. Chem.: Macromol. Symp. 12: 229
235. Gagnon DR, Capistran JD, Karasz FE, Lenz RW (1984) Polymer Prepr. 25(2): 282; (1984) Polym. Bull. 12: 293; (1985) Mol. Cryst. Liq. Cryst. 118: 327
236. Murase I, Ohnishi T, Noguchi T, Hirooka M (1984) Polymer Commun. 25: 327; (1985) Mol. Cryst. Liq. Cryst. 118: 333
237. Wessling RA, Zimmermann RG (1968) US Patent 3,401,152; (1972) US Patent 3,706,677
238. Kanbe M, Okawara M (1968) J. Polymer Sci. A1: 1058
239. Gagnon DR, Capistran JD, Karasz FE, Lenz RW, Antoun S (1987) Polymer 28: 567
240. Murase I, Ohnishi T, Noguchi T, Hirooka M (1985) Polymer Commun. 26: 362; (1987) Synth. Met. 17: 639
241. Antoun S, Gagnon DR, Karasz FE, Lenz RW (1986) Polymer Prepr. 27(1): 116; (1986) Polym. Bull. 15: 181; (1986) J. Polymer Sci.: Lett. 24: 503
242. Han C-C, Lenz RW, Karasz FE (1987) Polymer Commun. 28: 261
243. Antoun S, Gagnon DR, Karasz FE, Lenz RW (1986) J. Polymer Sci.: Lett. 24: 503
244. Murase I, Ohinishi T, Noguchi T, Hirooka M (1987) Polymer Commun. 28: 229
245. Yamada S, Tokito S, Tsutsui T, Saito S (1987) J. Chem. Soc.: Chem. Commun. 1448
246. Jen K, Maxfield M, Shacklette LW, Elsenbaumer RL (1987) J. Chem. Soc.: Chem. Commun.: 309
247. Marvel CS, Hartzell GE (1959) J. Amer. Chem. Soc. 81: 449
248. LeFebre G, Dawans F (1964) J. Polymer Sci.: A2: 3277
249. Cassidy PE, Marvel CS, Ray S (1965) J. Polymer Sci. 3: 1553
250. Ballard DGH, Courtis A, Shirley IM, Taylor SC (1983) J. Chem. Soc.: Chem. Commun.: 954
251. Stille JK, McKean DR (1987) Polymer Preprints 28(1): 65
252. Ballard DGH, Courtis A, Shirley IM, Taylor SC (1988) Macromolecules 21: 294
253. Edwards J, Fisher R, Vincent B (1983) Makromol. Chem.: Rapid Commun. 4: 393
254. Armes SP, Vincent B (1987) J. Chem. Soc.: Chem. Commun.: 288
255. Yassar A, Roncali J, Garnier F (1987) Polymer Commun. 28: 103
256. Soga K, Kobayashi Y, Ikeda S, Kawakami S (1982) J. Chem. Soc.: Chem. Commun.: 931
257. Aldissi M, Liepins R (1984) J. Chem. Soc.: Chem. Commun.: 255; (1985) Aldissi M, Poly. Prepr. 26(2): 269
258. Frommer JE (1986) Acc. Chem. Res. 19: 2
259. Frommer JE (1987) J. Macromol. Sci.: Chem. A24(3,4): 449
260. Sato M, Tanaka S, Kaeriyama K (1986) J. Chem. Soc.: Chem. Commun.: 873
261. Jen K-Y, Miller GG, Elsenbaumer RL (1986) J. Chem. Soc.: Chem. Commun.: 1347
262. Nowak MJ, Rughooputh SDDV, Hotta S, Heeger AJ (1987) Macromolecules 20: 965
263. Yoshino K, Nakajima S, Fujii M, Sugimoto R (1987) Polymer Commun 28: 309
264. Kaeriyama K, Sato M, Tanaka S (1987) Synth. Met. 18: 233
265. Elsenbaumer RL, Jen K-Y, Miller GG, Shacklette LW (1987) Synth. Met. 18: 277
266. Rughooputh SDDV, Hotta S, Heeger AJ, Wudl F (1987) J. Polymer Sci.: Phys. 25: 1071
267. Sato M, Tanaka S, Kaeriyama K (1987) Synth. Met. 18: 229
268. Bryce MR, Chissel A, Kathirgamanathan P, Parker D, Smith NRM (1987) J. Chem. Soc.: Chem. Commun. 466
269. Patil AO, Ikenoue Y, Basescu N, Colaneri N, Chen J, Wudl F, Heeger AJ (1987) Synth. Met. 20: 151
270. Havinga EE, van Horssen LW, ten Hoeve W, Wynberg H, Meijer EW (1987) Polymer Bull. 18: 277
271. Hotta S, Rughooputh SDDV, Heeger AJ (1987) Synth. Met. 22: 79
272. Li S, Cao Y, Xue Z (1987) Synth. Met. 20: 141
273. Davenport DE (1981) In: Seymour RB (Ed) Conductive polymers, Polymer Sci. Tech. 25: 39, Plenum, New York
274. Galvin ME, Wnek GE (1982) Polymer 23: 795; (1983) J. Polymer Sci.: Chem. 21: 2727
275. Sichel EK, Rubner MF (1985) J. Polymer Sci.: Phys. 23: 1629
276. Rubner MF, Tripathy SK, George J Jr, Cholewa P (1983) Macromolecules 16: 870
277. Tripathy SK, Rubner MF (1984) Amer. Chem. Soc. Symp. Ser. 242: 38
278. Lee KI, Jopson H (1983) Makromol. Chem.: Rapid Commun. 4: 375

279. Wessling B, Volk H (1987) Synth. Met. *18*: 671
280. Bates FS, Baker GL (1983) Macromolecules *16*: 704, 1013
281. Bates FS, Baker GL (1984) Macromolecules *17*: 2619
282. Aldissi M (1984) J. Chem. Soc.: Chem. Commun.: 1347
283. Aldissi M (1986) Synth. Met. *15*: 141
284. Aldissi M, Hou M, Farrell J (1987) Synth. Met. *17*: 229
285. Aldissi M, Bishop AR (1985) Polymer *26*: 622
286. Stowell JA, Amass AJ, Beevers MS, Farren TR (1987) Makromol. Chem. *188*: 1635
287. Farren TR, Amass AJ, Beevers MS, Stowell JA (1987) Makromol. Chem. *188*: 2535
288. Armes SP, Vincent B, White JW (1986) J. Chem. Soc.: Chem. Commun.: 1525
289. Destri S, Catellani M, Bolognesi A (1984) Makromol. Chem.: Rapid Commun. *15*: 353
290. Bolognesi A, Catellani M, Destri S, Porzio W, Meille SV, Pedemonte E (1986) Makromol. Chem. *187*: 1287
291. Cuniberti C, Fuso S, Piaggio P, Dellepiane G, Tubino R (1987) Synth. Met. *21*: 313
292. Cuniberti C, Piaggio P, Dellepiane G, Catellani M, Piseri L, Porzio W, Tubino R (1986) Makromol. Chem.: Rapid Commun. 7: 471
293. VanNice FL, Bates FS, Baker GL, Carroll PJ, Patterson GD (1984) Macromolecules *17*: 2626
294. Piaggio P, Cuniberti C, Dellepiane G, Bolognesi A, Catellani M, Destri S, Porzio W, Tubino R (1987) Synth. Met. *17*, 337
295. Ahlgren G, Krische B (1984) J. Chem. Soc.: Chem. Commun.: 946
296. DePaoli M-A, Waltman RJ, Diaz AF, Bargon J (1984) J. Chem. Soc.: Chem. Commun.: 1015
297. Niwa O, Tamamura T (1984) J. Chem. Soc.: Chem. Commun.: 817
298. Niwa O, Tamamura T (1985) Makromol. Chem.: Rapid Commun. *16*: 375
299. Niwa O, Kakuchi M, Tamamura T (1987) Macromolecules *20*: 749
300. Niwa O, Hikita M, Tamamura T (1987) Synth. Met. *18*: 677
301. Lindsey SE, Street GB (1984/5) Synth. Met. *10*: 67
302. Niwa O, Kakuchi M, Tamamura T (1987) Polymer J. *19*: 1293
303. Iyoda T, Ohtani AH, Shimidzu T, Honda K (1986) Chem. Lett.: 687
304. Shimidzu T, Ohtani AH, Iyoda T, Honda K (1986) J. Chem. Soc. Chem. Commun.: 1415
305. Bates N, Cross M, Lines R, Walton D (1985) J. Chem. Soc. Chem. Commun.: 871
306. Penner RM, Martin CR (1986) J. Electrochem. Soc. *33*: 310
307. Roncali J, Garnier F (1986) J. Chem. Soc.: Chem. Commun.: 783
308. Jasne SJ, Chiklis CK (1986) Synth. Met. *15*: 175
309. Bocchi V, Gardini GP (1986) J. Chem. Soc.: Chem. Commun.: 148
310. Bocchi V, Gardini GP, Rapi S (1987) J. Mater. Sci. Lett. *6*: 1283
311. Aldebert P, Audebert P, Armand M, Bidan G, Pineri M (1986) J. Chem. Soc.: Chem. Commun.: 1636
312. Bjorklund RB, Liedberg B (1986) J. Chem. Soc.: Chem. Commun.: 1293
313. Ojio T, Miyata S (1986) Polym. J. *18*: 95
314. Yosomiya R, Hirata M, Haga Y, An H, Seki M (1986) Makromol. Chem.: Rapid Commun. 7: 697
315. Huang W-S, Park JM (1987) J. Chem. Soc.: Chem. Commun.: 856
316. Aizawa M, Yamada T, Shinohara H, Akagi K, Shirakawa H (1986) J. Chem. Soc.: Chem. Commun.: 1315
317. Machado JM, Schlenhoff JB, Glatzowski PJ, Karasz FE (1987) Polym. Mater. Sci. Eng. *57*: 446
318. Graessley WW (1984) In: Physical properties of polymers, Amer. Chem. Soc. Washington
319. Billingham NC (1977) Molar mass measurements in polymer science, Kogan Page, London
320. Shirakawa H, Sato M, Hamano A, Kawakami S, Soga K, Ikeda S (1980) Macromolecules *13*: 457
321. Enkelmann V, Leiser G, Monkenbusch M, Muller W, Wegner G (1981) Mol. Cryst. Liq. Cryst. *77*: 111
322. Schen MA, Karasz FE, Chien JCW (1983) J. Polymer Sci.: Chem. *21*: 2787
323. Chien JCW, Babu GN (1985) Macromolecules *18*: 622
324. Chien JCW, Schen MA (1985) J. Polymer Sci. *23*: 2447
325. Schen MA, Karasz FE, Chien JCW (1984) Makromol. Chem. *5*: 217

326. Harper K, James PG (1985) Mol. Cryst. Liq. Cryst. *117*: 55
327. Naarman H (1979) Ber. Bunsenges. Phys. Chem. *83*: 427
328. Jones MB, Kovacik P, Lanska D (1981) J. Polymer Sci.: Chem. *19*: 89
329. Brown CE, Jones MB, Kovacik P (1980) J. Polymer Sci.: Lett. *18*: 653
330. Brown CE, Khoury I, Bezoari MD, Kovacik P (1982) J. Polymer. Sci.: Chem. *20*: 1697
331. Durham JE, Kovacik P (1977) J. Polym. Sci.: Chem. *15*: 2701
332. Jones MB, Kovacik P, Lanska D (1981) J. Polymer Sci.: Chem. *19*: 89
333. Brown CE, Kovacik P, Wilkie CA, Cody RB, Kinsinger JA (1985) J. Polymer Sci.: Lett. *23*: 453
335. Havinga EE, van Horssen LW (1987) Synth. Met. *17*: 623
335. Cao Y, Wu Q, Guo K, Quian R (1984) Makromol. Chem. *185*: 389
336. Nazzal A, Street GB (1984) J. Chem. Soc.: Chem. Commun.: 83
337. Hotta S, Rughooputh SDDV, Heeger AJ, Wudl F (1987) Macromolecules *20*: 212
338. Watanabe A, Mori K, Iwaski Y, Nakamura Y (1987) J. Chem. Soc.: Chem. Commun.: 3
339. Stacey CJ (1986) J. Appl. Polymer Sci. *32*: 3959
340. Li GCG, Kidwell DA, Brown DA, Wall EE, Wnek GE (1982) J. Polymer Sci.: Chem. *21*: 301
341. Gibson HW, Pochan JM, Kaplan S (1981) J. Amer. Chem. Soc. *103*: 4619
342. Clarke TC, Krounby MT, Lee VY, Street GB (1981) J. Chem. Soc.; Chem. Commun.: 384
343. Haberkorn H, Naarman H, Peuzien K, Schlag K, Simak P (1982) Synth. Met. *5*: 51
344. a) Tanaka K, Yamanaka S, Oji M, Yamabe T (1988) Synth. Met. *22*: 247; b) (1984) Terao T, Maeda S, Yamabe T, Akagi K, Shirakawa H, Chem. Phys. Lett. *13*: 347
345. Gibson HW, Weagley RJ, Prest WM Jr, Mosher R, Kaplan S (1983) J. de Physique *C3*: 123
346. Akaishi T, Miyasala K, Ishiawa K, Shirakawa H, Ikeda S (1980) J. Polymer Sci.: Phys. *18*: 745
347. Foot PJS, Billingham NC, Calvert PD, Brown CS, Walker NI, James DI (1986) Polymer *27*: 448
348. Brown CE, Kovacik P, Wilkie CA, Kinsinger JA, Hein RE, Yaniger SI, Cody RB (1986) J. Polymer Sci.: Chem. *24*: 255
349. Duroux JL, Moliton A, Froyer G, Maurice F (1986) Eur. Polym. J. *22*: 439
350. Diaz AF, Martinez M, Kanawa KK, Salmon M (1981) J. Electroanal. Chem. Interfacial Electrochem. *130*: 181
351. Osterholm J-E, Sunila P, Hjertberg T (1987) Synth. Met. *18*: 169
352. Brown CE, Kovacik P, Cody RB, Hein RE, Kinsinger JA (1986) J. Polymer Sci.: Lett. *24*: 519
353. Chanzy HD, Fisa B, Marchessault RH (1973) Crit. Rev. Macromol. Sci. *1*: 315
354. Wunderlich B (1973) Macromolecular physics, Academic, New York, vol 1, p 341
355. Abadie MJM, Hacene SMB, Cadene M, Rolland M (1986) Polymer *27*: 2003
356. Davenas J, Xu XL, Maitrot M, Francois B, Mathis C, Andre JJ (1987) Mol. Cryst. Liq. Cryst. *117*: 87
357. Wang F, Zhao X, Gong Z, Cao Y, Yan Q, Qian R (1982) Makromol. Chem. Rapid Commun. *3*: 929
358. Lieser G, Wegner G, Muller W, Enkelmann V (1980) Makromol. Chem.: Rapid Commun. *1*: 621
359. Karasz FE, Chien JCW, Galkiewicz R, Wnek GE, Heeger AG, MacDiarmid AG (1979) Nature *282*: 236
360. Chien JCW, Karasz FE, Schen MA, Yamashita Y (1983) Makromol. Chem.: Rapid Commun. *4*: 5
361. Pennings AJ, van der Mark JMAA, Keil AM (1970) Koll. Z. Z. Polym. *237*: 336
362. Haberkorn H, Naarmann H, Penzien K, Schlag J, Simak P (1982) Synth. Met. *15*: 51
363. Chien JCW, Yamashita Y, Hirsch JA, Fan JL, Schen MA, Karasz FA (1982) Nature *299*: 608
364. Druy MA, Tsang CH, Brown N, Heeger AJ, MacDiarmid AG (1980) J. Polymer Sci.: Phys. *18*: 429
365. Masuda T, Tang B-Z, Tanaka A, Higashimura T (1986) Macromolecules *19*: 1459
366. Meyer WH (1981) Synth. Met. *4*: 81; (1981) Mol. Cryst. Liq. Cryst. *77*: 137
367. Aldissi M (1985) J. Polymer Sci.: Lett. *23*: 167

368. Akagi K, Katayama S, Shirakawa H, Araya K, Mukoh A, Narahara T (1987) Synth. Met. *17*: 241
369. Araya K, Mukoh A, Narahara T, Akagi K, Shirakawa H (1986) Synth. Met. *14*: 199
370. Araya K, Mukoh A, Narahara T, Akagi K, Shirakawa H (1987) Synth. Met. *17*: 247
371. Akagi K, Shirakawa H, Araya K, Mukoh A, Narahara T (1987) Synth. Met. *19*: 185
372. Montaner A, Rolland M, Sauvajol JL, Galtier M, Almairac R, Ribet JL (1988) Polymer *29*: 1101
373. Woerner T, MacDiarmid AG, Heeger AJ (1982) J. Polym. Sci.: Lett. *20*: 305
374. Yamashita Y, Shimamura K, Kasahara H, Monobe K (1987) Synth. Met. *17*: 253
375. Yamashita Y, Nishimura S, Shimamura K, Monobe K (1986) Makromol. Chem. *187*: 1757
376. Fincher CR, Peebles DL, Heeger AJ, Druy MA, Matsumura Y, MacDiarmid AG, Shirakawa H, Ikeda S (1978) Solid State Commun. *27*: 489
377. Lugli G, Pedretti U, Perego G (1985) J. Polymer Sci.: Lett. *23*: 129
378. White D, Bott DC, Whitehead RH (1983) Polymer *24*: 805
379. Kahlert H, Leising G (1985) Mol. Cryst. Liq. Cryst. *117*: 1; Liesing G (1984) Polymer. Commun. *25*: 201
380. Schonhorn H, Baker GL, Bates FS (1985) J. Polymer Sci.: Phys. *23*: 1555
381. Giuseppi-Elie A, Wnek GE (1985) J. Polymer Sci.: Chem. *23*: 2601
382. Epstein AJ, Rommelmann H, Fernquist R, Gibson HW, Druy MA (1982) Polymer *23*: 1211; Rommelmann H, Fernquist R, Epstein AJ, Aldissi M, Woerner T, Bernier P (1983) Polymer *24*: 1575
383. Begin D, Demai JJ, Vangelisti R, Billaud D (1986) Polymer Commun. *27*: 117
384. Rolland M, Aldissi M, Schue F (1982) Polymer *23*: 835
385. Rachdi F, Bernier P, Falques E, Lefrant S, Schue F (1982) Polymer *23*: 173
386. Elsenbaumer RL, Shacklette LW (1986) In: Skotheim TA (ed), Handbook of conducting polymers, Dekker, New York, vol 1, chap 7
387. Teraoka I, Takahashi T (1980) J. Macromol. Sci.: Phys. *18*: 73
388. Lewis IC, Kovac CA (1979) Mol. Cryst. Liq. Cryst. *51*: 173; (1979) Mol. Cryst. Liq. Cryst. *49*: 207
389. Pradere P, Boudet A, Goblot J-Y, Froyer G, Maurice F (1985) Mol. Cryst. Liq. Cryst. *118*: 277
390. Froyer G, Maurice F, Mercier JP, Riviere D, LeCun M, Auvray P (1981) Polymer *22*: 992
391. Gale DM (1978) J. Appl. Polymer Sci. *22*: 1955, 1971
392. Diaz A (1981) Chem. Script. *117*: 145
393. Street GB (1986) In: Skotheim TA (ed) Handbook of conducting polymers, Dekker, New York, vol 1, chap 8
394. Ogasawara M, Funahashi K, Iwata K (1985) Mol. Cryst. Liq. Cryst. *118*: 159
395. Bloor D, Hercliffe RD, Galiotis CG, Young RJ (1985) In: Kuzmany H, Mehring M, Roth S (eds) Electronic properties of polymers and related compounds, Springer, Berlin, Heidelberg, New York
396. Naarmann H (1987) Makromol. Chem. Symp. *8*: 1
397. Buckley LJ, Roylance DK, Wnek GE (1987) J. Polymer Sci.: Phys. *25*: 2179
398. Tourillon G, Garnier F (1984) J. Polymer Sci.: Phys. *22*: 33
399. Ito M, Shioda H, Tanaka K (1986) J. Polymer Sci.: Lett. *24*: 147
400. Satoh M, Yamasaki H, Aoki S, Yoshino K (1987) Polymer Commun. *28*: 144
401. Ito M, Tsurunto A, Osawa S, Tanaka K (1988) Polymer *29*: 1161
402. Wang B, Tang J, Wang F (1986) Synth. Met. *13*: 329
403. Wang B, Tang J, Wang F (1987) Synth. Met. *18*: 323
404. Chen S-A, Lee T-S (1987) J. Polymer Sci. Lett. *25*: 455
405. Kitani A, Kaya M, Tsujioka S, Sasaki K (1988) J. Polymer Sci.: Chem. *26*: 1531
406. Wunderlich B (1976) Macromolecular physics, Academic, New York, vol 2
407. Capaccio G, Ward IM (1974) Polymer *15*: 233
408. Smith P, Lemstra PJ (1980) J. Mater. Sci. *15*: 505
409. See e.g. Bredas JL (1986) In: Skotheim TA (ed) Handbook of conducting polymers, Dekker, New York, vol 2, Ch 25
410. See e.g. Bredas JL, Sibley R, Boudreaux DS, Chance RR (1983) J. Amer. Chem. Soc. *105*: 6555

411. Bredas JL (1987) Synth. Met. *17*: 115
412. Jenekhe SA (1986) Nature *322*: 345
413. MacDiarmid AG, Mammone RJ, Crawczyk JR, Porter SJ (1984) Mol. Cryst. Liq. Cryst. *105*: 89
414. Friend RH, Bradley DDC, Townsend PD (1987) J. Phys.: D, *20*: 1387
415. Gerrard DL, Maddams WF (1981) Macromolecules *14*: 1356
416. Rimai L, Heyde ME, Gill D (1973) J. Amer. Chem. Soc. *95*: 4493
417. Kuzmany H, Knoll P (1985) Mol. Cryst. Liq. Cryst. *117*: 385
418. Williams KPJ, Gerrard DL, Bott DC, Chai CK (1985) Mol. Cryst. Liq. Cryst. *117*: 235
419. Foot PJS, Billingham NC, Calvert PD (1986) Synth. Met. *16*: 265
420. Bott DC, Brown CS, Winter JN, Barker J (1987) Polymer *28*: 601
421. Bowley HJ, Gerrard DL, Maddams WF (1987) Makromol. Chem. *188*: 899
422. Baughman RH, Shacklette LW (1987) Synth. Met. *17*: 173
423. Friend RH, Bradley DDC, Pereira CM, Townsend PD, Bott DC, Williams KPJ (1985) Synth. Met. *13*: 101
424. Chien JCW, Babu GN, Hirsch JA (1985) Nature *314*: 723
425. Chien JCW, Babu GN (1985) J. Chem. Phys. *82*: 441
426. Wudl F, Heeger AJ, Frommer JE (1986) Nature *319*: 697
427. Chien JCW (1986) Nature *319*: 698
428. Schafer-Siebert D, Roth S, Budrowski C, Kuzmany H (1987) Synth. Met. *21*: 285
429. Yang XQ, Tanner DB, Arbuckle G, MacDiarmid AG, Epstein AJ (1987) Synth. Met. *17*: 277
430. Abadie MJM, Djebaili A, Cadene M, Rolland M (1988) Eur. Polymer J. *24*: 251
431. See e.g. van Krevelen DA (1972) Properties of polymers, Elsevier, New York
432. Perego G, Lugli G, Pedretti V (1985) Mol. Cryst. Liq. Cryst. *117*: 59
433. Robin P, Pouget JP, Comus R, Gibson HW, Epstein AJ (1983) J. de Phys. *44*: C3–77
434. Shacklette LW, Elsenbaumer RL, Chance RR, Eckhardt H, Frommer JE, Baughman RH (1981) J. Chem. Phys. *75*: 1919
435. Natta G, Corradini P (1960) Nuov. Cim. Supp. *15*: 9
436. Wunderlich B (1973) Macromolecular physics, Academic, New York, vol 1, chap 2
437. Karpfen A, Holler R (1981) Solid State Commun. *37*: 179
438. Yamabe T, Kanaka K, Terame H, Fukui H, Immamura A, Shirakawa H, Ikeda S (1979) Solid State Commun. *29*: 329
439. Baughman RH, Hsu SL, Petz GP, Signorelli AJ (1978) J. Chem. Phys. *68*: 5405
440. Chien JCW, Karasz FE, Shimamura K (1982) J. Polymer Sci.: Lett. *20*: 97
441. Bates FS, Baker GL (1983) Macromolecules *16*: 1013
442. Elert ML, White CT, Mintmire JW (1985) Mol. Cryst. Liq. Cryst. *125*: 329
443. Elert ML, White CT (1987) Macromolecules *20*: 1411
444. Fincher CR, Chen CE, Heeger A, MacDiarmid AG, Hastings JB (1982) Phys. Rev. Lett. *48*: 100
445. Brown CS, Vickers ME, Foot PJS, Billingham NC, Calvert PD (1986) Polymer *27*: 1719
446. Gibson HW, Prest WM, Mosher RA, Kaplan S, Weagley RA (1983) Polymer Prep. *24(2)*: 153
447. Chien JCW, Yang X (1983) J. Polymer Sci.: Lett. *21*: 767
448. Chien JCW (1981) J. Polymer Sci.: Lett. *19*: 249
449. Hsu SL, Signorelli AJ, Pez GP, Baughman RH (1978) J. Chem. Phys. *69*: 106
450. Shimamura K, Karasz FE, Chien JCW, Hirsch JA (1982) Makromol. Chem.: Rapid Commun. *3*: 269
451. Monkenbusch M, Morra BS, Wegner G (1982) Makromol. Chem.: Rapid Commun. *3*: 69, 601
452. Matsuyama T, Sakai H, Yamaoda H, Maeda Y, Shirakawa H (1981) Solid State Commun. *40*: 563
453. Kaindl G, Wortmann G, Roth S, Menke K (1982) Solid State Commun. *41*: 75
454. Lefrant S, Lichtmann LS, Temkin H, Fitchen DB, Miller PC, Whitwell GE, Burlitch JM (1979) Solid State Commun. *29*: 191
455. Baughman RH, Murthy NS, Miller GG, Shacklette LW (1983) J. Chem. Phys. *79*: 1065
456. Petit MA, Soum AH, Leclerc M, Prud'homme RE (1987) J. Polymer Sci.: Phys. *25*: 423
457. Vogel FL, Foley GMT, Zeller C, Falardeau ER, Gan JS (1977) Mater. Sci. Eng. *31*: 261
458. Clarke TC, Street GB (1979) Synth. Met. *1*: 119

459. Krone W, Wortmann G, Frank KH, Godler F, Kaindl G, Menke K, Weizenhofer R (1987) Synth. Met. *17*: 383
460. Hasslin HW, Riekel C, Menke K, Roth S (1984) Makromol. Chem. *185*: 397
461. Riekel C, Hasslin HW, Menke K, Roth S (1984/5) Synth. Met. *10*: 31
462. Begin D, Demai JJ, Vangelisti R, Billaud D (1986) Polymer Commun. *27*: 97
463. Pouget JP, Pouxviel JC, Robin P, Comes R, Begin D, Billaud DA, Feldblum A, Gibson HW, Epstein AJ (1985) Mol. Cryst. Liq. Cryst. *117*: 75
464. Flandrois S, Hauw C, Francois B (1985) Mol. Cryst. Liq. Cryst. *117*: 91
465. Shacklette LW, Murthy NS, Baughman RH (1985) Mol. Cryst. Liq. Cryst. *121*: 201
466. Baughman RH, Murthy NS, Miller GG (1983) J. Chem. Phys. *79*: 515
467. Baughman RH, Shacklette LW, Murthy NS, Miller GG, Elsenbaumer RL (1985) Mol. Cryst. Liq. Cryst. *118*: 253
468. Kahlert H, Leitner O, Leising G (1987) Synth. Met. *17*: 467
469. Sokolowski MM, Marseglia EA, Friend RH (1986) Polymer *27*: 1714
470. Bradley DDC, Friend RH, Hartmann T, Marseglia EA, Sokolowski MM, Townsend PD (1987) Synth. Met. *17*: 473
471. Moon YB, Winokur M, Heeger AJ, Barker J, Bott DC (1987) Macromolecules *20*: 2457
472. Kovacic P, Feldman MB, Kovacic JP, Lando JB (1968) J. Appl. Polymer Sci. *12*: 1735
473. Teraoka I, Takahashi T (1980) J. Macromol. Sci. Phys. *B18*: 73
474. Hasslin HW, Riekel C (1982) Synth. Met. *5*: 37
475. Pradere P, Boudet A, Goblot J-Y, Froyer G, Maurice F (1985) Mol. Cryst. Liq. Cryst. *118*: 277
476. Bolognesi A, Catellani M, Destri S, Porzio W (1985) Polymer *26*: 1628
477. Barbarin F, Berthet G, Blanc JP, Fabre C, Germain JP, Handi M, Robert H (1983) Synth. Met. *16*: 53
478. Czerwinski W, Bala W, Kreja L (1985) Angew. Makromol. Chem. *123*: 1132
479. Stamm M, Fink J, Tieke B (1985) Mol. Cryst. Liq. Cryst. *118*: 281
480. Enkelmann V, Wieners G, Eiffler J (1983) Makromol. Chem.: Rapid Commun. *4*: 337
481. Tieke B, Bubeck C, Lieser C (1982) Makromol. Chem.: Rapid Commun. *3*: 261
482. Stamm M (1984) Mol. Cryst. Liq. Cryst. *105*: 259
483. Goblot J-Y, Maurice F, Froyer G, Pelous Y (1985) Mol. Cryst. Liq. Cryst. *118*: 273
484. Mitchell GR (1986) Polymer Commun. *27*: 346
485. Pfluger P, Krounbi M, Street GB, Weiser G (1983) J. Chem. Phys. *78*: 3212
486. Pfluger P, Street GB (1985) J. Chem. Phys. *80*: 544
487. Salaneck WR, Erlandsson R, Prejza J, Lundstrom I, Inganas O (1983) Synth. Met. *5*: 125
488. Street GB, Lindsey SE, Nazzal AI, Wynne KJ (1985) Mol. Cryst. Liq. Cryst. *118*: 137
489. Geiss RH, Street GB, Volksen W, Economy J (1983) IBM J. Res. Dev. *27*: 321
490. Wegner G (1984) Mol. Cryst. Liq. Cryst. *106*: 269
491. Mo Z, Lee K-B, Moon YB, Kobayashi M, Heeger AG, Wudl F (1988) Macromolecules *18*: 1972
492. Bruckner S, Porzio W (1988) Makromol. Chem. *189*: 961
493. Tourillon G, Garnier F (1985) Mol. Cryst. Liq. Cryst. *121*: 349
494. Gagnon DR, Kapistran JD, Karasz FE, Lenz RW (1984) Polym. Bull. *12*: 293
495. Murase I, Ohnishi T, Noguchi T, Hirooka M (1984) Polymer *25*: 327
496. Granier T, Thomas EL, Gagnon DR, Karasz FE, Lenz RW (1986) J. Polymer Sci.: Phys. *24*: 2793
497. Bradley DDC, Friend RH, Lindenberger H, Roth S (1986) Polymer *27*: 1709
498. Bradley DDC (1987) J. Phys.: D *20*: 1389
499. Masse MA, Martin DC, Karasz FE, Thomas EL (1987) Polym. Eng. Sci. *57*: 441
500. Panar M, Avakian P, Blume RC, Gardner KH, Gierke TD, Yang HH (1983) J. Polymer Sci.: Phys. *21*: 1955
501. Crank J (1975) The mathematics of diffusion, 2nd edn, Clarendon, Oxford
502. Alfrey T, Gurney EF, Lloyd WG (1966) J. Polymer Sci. *C12*: 249
503. Thomas NL, Windle AH (1980) Polymer *21*: 613; (1982) *23*: 529
504. Danno T, Miyasaka K, Ishikawa K (1983) J. Polymer Sci.: Phys. *21*: 1527
505. Beniere F, Haridoss H, Louboutin JP, Aldissi H, Fabre JM (1981) J. Phys. Chem. Solids *42*: 649
506. Pekker S, Bellec M, LeCleac'h X, Beniere F (1984) Synth. Met. *9*: 475

507. Radici R, Bernier F, Falques E, Lefrant S, Schue F (1983) J. Phys. *C44*: 97
508. Kaner RB, MacDiarmid AG (1984) J. Chem. Soc. Faraday Trans. I, *80*: 2109
509. Armand M (1983) J. de Physique *44*: C3-551
510. Foot PJS, Nevett BA (1983) Solid State Ionics *8*: 169
511. Will FG (1985) J. Electrochem. Soc. *132*: 2093
512. Francois B, Mathis C, Nuffer R (1987) Synth. Met. *20*: 311
513. Chen S-A, Chan W-C, Li L-S (1987) Angew. Makromol. Chem. *148*: 87
514. Foot PJS, Calvert PD, Ware M, Billingham NC, Bott DC (1985) Mol. Cryst. Liq. Cryst. *117*: 47
515. Foot PJS, Mohammed F, Calvert PD, Billingham NC (1987) J. Phys. *D20*: 1354
516. Clarke TC, Geiss RH, Gill WD, Grant VM, Morawitz H, Street GB (1979) Synth. Met. *1*: 21
517. Stannett V (1968) In: Crank J, Park GS (eds) Diffusion in polymers, Academic, London
518. Mirebeau P (1983) J. de Physique *44*: C3-579
519. Genies EM, Pernaut JM (1984/5) Synth. Met. *10*: 117
520. Kaufman JH, Kanazawa KK, Street GB (1984) Phys. Rev. Lett. *53*: 2461
521. Shinohara H, Aizawa M, Shirakawa H (1986) J. Chem. Soc.; Chem. Commun.: 87
522. Tietje-Girault J, Anderson JM, MacInnes I, Schroder M, Tennant G, Girault HH (1987) J. Chem. Soc.; Chem. Commun.: 1095
523. Kaneto K, Agawa H, Yoshino K (1987) J. Appl. Phys. *61*: 1197
524. Kaneto T, Yoshino K, Inuishi Y (1983) Jap. J. Appl. Phys. *22*: L412
525. Grassie N, Scott G (1985) Polymer degradation and stability, Cambridge Univ. Press
526. Ito T, Shirakawa H, Ikeda S (1975) J. Polymer Sci.: Chem. *13*: 1943
527. Chien JCW, Uden PC, Fan J-L (1982) J. Polymer Sci.: Chem. *20*: 2159
528. Fan J-L, Chien JCW (1983) J. Polymer Sci.: Chem. *21*: 3453
529. Cukor P, Rubner M (1980) J. Polymer Sci.: Phys. *18*: 909
530. Druy MA, Rubner MF, Walsh SP (1986) Synth. Met. *13*: 207
531. Hatano M, Kambara S, Okamoto S (1961) J. Polymer Sci. *51*: 26
532. Shirakawa H, Ikeda S (1971) Polymer J. *2*: 231
533. Chang CK, Park YW, Heeger AJ, Shirakawa H, Louis EJ, MacDiarmid AJ (1978) J. Chem. Phys. *69*: 5098
534. Goldberg JB, Crowe HR, Newman PR, Heeger AJ, MacDiarmid AG (1979) J. Chem. Phys. *70*: 1132
535. Snow A, Brant P, Weber D, Yang NL (1979) J. Polymer Sci.: Lett. *17*: 263
536. Vansco G, Rockenbauer A (1981) Chem. Scrip. *17*: 153
537. Chien JCW, Karasz FE, Wnek GE, MacDiarmid AJ, Heeger AJ (1980) J. Polymer Sci.: Lett. *18*: 45
538. Pochan JM (1986) In: Skotheim TA (ed) Handbook of conducting polymers, Dekker, vol 2, chap 39
539. Pochan JM, Gibson HW, Bailey FC (1980) J. Polymer Sci.: Lett. *18*: 447
540. Pochan JM, Gibson HW, Bailey FC, Pochan DF (1980) Polymer *21*: 250
541. Yen SPS, Somoano R, Khanna SK, Rembaum A (1980) Solid State Commun. *36*: 339
542. Helmie M, Becker J, Mehring M (1986) In: Kuzmany H, Mehring M, Roth S (eds) Electronic properties of polymers and related compounds, Berlin Heidelberg New York, p. 275
543. Sohma J, Tabata M, Ebisawa F, Hatano M (1987) Polymer Deg. Stab. *17*: 5
544. Higashimura T, Tang B-Z, Masuda T, Yamaoka H, Matsuyama T (1985) Polymer J. *17*: 393
545. Pochan JM, Pochan DF, Rommelmann H, Gibson HW (1981) Macromolecules *14*: 110
546. Gibson HW, Pochan JM (1982) Macromolecules *15*: 242
547. Turro NJ (1967) Molecular photochemistry, Benjamin, New York, p 176
548. Bernier P, Rolland M, Linaya C, Aldissi M (1980) Polymer Commun. *21*: 7
549. Lefrant S, Rzepka E, Bernier P, Rolland M, Aldissi M (1980) Polymer Commun. *21*: 1235
550. Chien JCW, Yang X (1983) J. Polymer Sci.: Lett. *21*: 767
551. Ito T, Shirakawa H, Ikeda S (1975) J. Polymer Sci.: Chem. *13*: 1943
552. Zanobi A, D'Ilario L (1983) J. de Physique *C3*: 309
553. Cernia E, D'Ilario L, Lupoli M, Mantovani E, Schwarz M, Benni P, Zanobi A (1984) J. Polymer Sci.: Chem. *22*: 3393
554. Haleem MA, Billaud D, Pron A (1982) Polymer *23*: 1409
555. Haleem MA, Chenite A, Billaud D (1983) Polymer *24*: 54

556. Huq R, Farrington GC (1984) J. Electrochem. Soc. *131*: 819
557. Ebisawa F, Tabei H (1985) J. Appl. Phys. *58*: 2326
558. Kivelson S (1981) Phys. Rev. Lett. *46*: 1344
559. Will FG, McKee DW (1983) J. Polymer Sci.: Chem. *21*: 3479
560. Chien JCW, Dickinson LC, Yang X (1983) Macromolecules *16*: 1287
561. Yang X-Z, Chien JCW (1985) J. Polymer Sci.: Chem. *23*: 859
562. Druy MA, Tsang CH, Brown N, Heeger AJ, MacDiarmid AG (1980) J. Polymer Sci.: Phys. *18*: 429
563. Chen S, Li L (1984) Makromol. Chem. *185*: 1063
564. Pochan JM, Gibson HW, Harbour J (1982) Polymer *23*: 435
565. Muller HK, Hocker J, Menke K, Ehinger H, Roth S (1985) Synth. Met. *10*: 273
566. Ohtsuka K, Nagata S, Akiya T (1987) Synth. Met. *17*: 289
567. Aldissi M (1984) Synth. Met. *9*: 131
568. Billingham NC, Calvert PD, Foot PJS, Mohammed F (1987) Polymer Deg. Stab. *19*: 323
569. Pron A, Budrowski C, Przyluski J (1983) Polymer *24*: 1294
570. Pron A, Falques E, Lefrant S (1987) Polymer Commun. *28*: 27
571. MacDiarmid AG, Mammone RJ, Krawczyk JR, Porter SJ (1984) Mol. Cryst. Liq. Cryst. *105*: 89
572. Mammone RJ, MacDiarmid AG (1984) Synth. Met. *9*: 143
573. Giuseppi-Elie A, Wnek GE (1983) J. Chem. Soc.: Chem. Commun.: 63
574. Wanqun W, Mammone RJ, MacDiarmid AG (1985) Synth. Met. *10*: 235
575. Terlemeyzan L, Mihailov M, Ivanova B (1985) Makromol. Chem.: Rapid Commun. *6*: 619
576. Mu S-L (1986) Synth. Met. *14*: 19
577. Will FG (1986) J. Electrochem. Soc. *132*: 518
578. Elsenbaumer RL, Delannoy P, Miller GG, Forbes CE, Murthy NS, Eckhardt H, Baughman RH (1985) Synth. Met. *11*: 251
579. Koga K, Kawakami S, Shirakawa H (1980) Makromol. Chem.: Rapid Commun. *1*: 643
580. Koga K, Nakamaru M (1983) J. Chem. Soc.: Chem. Commun.: 1495
581. Yaniger SI, Kletter MJ, MacDiarmid AG (1984) Polymer Prepr. *25(2)*: 264
582. Whitney DH, Wnek GE (1988) Macromolecules *21*: 266
583. Masuda T, Tang B-Z, Higashimura T (1985) Macromolecules *18*: 2369
584. Deits W, Cukor P, Rubner M, Jopson H (1982) Synth. Met. *4*: 199
585. Pochan JM, Pochan DF, Gibson HW (1981) Polymer *22*: 1367
586. Aldissi M (1986) Synth. Met. *15*: 141
587. Pfluger P, Krounbi M, Street GB, Weiser G (1983) J. Chem. Phys. *78*: 3212
588. Salanek WR, Erlandsson R, Prejza J, Lundstrom I, Inganas O (1983) Synth. Met. *5*: 125
589. Diaz AF, Kanazawa KK (1979) J. Chem. Soc.: Chem. Commun.: 854
590. Munstedt H (1986) In: Kuzmany H, Mehring M, Roth S (eds) Electronic properties of polymers and related compounds. Springer, Berlin Heidelberg New York, p 8
591. Erlandsson R, Inganas O, Lundstrom I, Salanek WR (1985) Synth. Met. *10*: 303
592. Hahn SJ, Gajda WJ, Vogelhut PO, Zeller MV (1986) Synth. Met. *14*: 89
593. Wernet W, Wegner G (1987) Makromol. Chem. *188*: 1465
594. Tourillon G, Garnier F (1983) J. Electrochem. Soc. *130*: 2043
595. Takenaka Y, Koike T, Oka T, Tanahashi M (1987) Synth. Met. *18*: 207
596. Corradini A, Mastrogostino M, Panero AS, Prosperi P, Scrosati B (1987) Synth. Met. *18*: 625
597. Munstedt H (1988) Polymer *29*: 296
598. Elsenbaumer RL, Maleysson C, Jen KY (1988) Polymer Mater. Sci. Eng. *56*: 54
599. MacDiarmid AG, Kaner RB (1986) In: Skotheim TA (ed) Handbook of conducting polymers. Dekker, New York, vol 1 chap 20
600. Kaner RB, MacDiarmid AG, Mammone RJ (1984) Amer. Chem. Soc. Symp. Ser. *242*: 575
601. Chiang CK (1981) Polymer *22*: 1454
602. Wegner G (1986) Makromol. Chem.: Symp. *1*: 151
603. Bittihn R (1985) Springer Solid State Sci *63*: 206
604. Passiniemi P, Osterholm J-E (1987) Synth. Met. *18*: 637
605. Nagatomo T, Ichikawa C, Omoto O (1987) J. Electrochem. Soc. *134*: 305; (1987) Synth. Met. *18*: 649

606. Wieners G, Monkenbusch M, Wegner G (1984) Ber. Bunsenges. Phys. Chem. *88*: 935
607. Farrington GC, Scrosati B, Frydrych D, DeNuzzio J (1984) J. Electrochem. Soc. *131*: 7
608. Shacklette LW, Elsenbaumer RL, Chance RR, Sowa JM, Ivory DM, Miller GG, Baughman RH (1982) J. Chem. Soc.: Chem. Commun.: 361
609. Dietrich M, Mortensen J, Heinze J (1986) J. Chem. Soc.: Chem. Commun.: 1131
610. Bittihn R, Ely G, Woeffler F, Munstedt H, Naarman H, Naegele D (1987) Makromol. Chem.; Symp. *8*: 51
611. Mohammadi A, Inganas O, Lundstrom I (1986) J. Electrochem. Soc. *133*: 947
612. Shimidzu T, Ohtani A, Iyoda T, Honda K (1987) J. Chem. Soc.; Chem. Commun.: 327
613. Kiani A, Kaya M, Asaki KS (1986) J. Electrochem. Soc. *133*: 1069
614. MacDiarmid AG, Mu S-L, Somasiri NLD, Wu W (1985) Mol. Cryst. Liq. Cryst. *121*: 187
615. Shacklette LW, Maxfield M, Gould S, Wolf JF, Jow TR, Baughman RH (1987) Synth. Met. *18*: 611
616. Kanicki J (1986) In: Skotheim (ed) Handbook of conducting polymers. Dekker, New York, vol 1 chap 17
617. Kaneko M, Yamada A, Kenmochi T, Tsuchida E (1985) J. Polymer Sci.: Lett. *23*: 629
618. Wada T, Takeno A, Iwaki M, Sasabe H, Kobayashi Y (1985) J. Chem. Soc.: Chem. Commun.: 1194
619. Aizawa M, Yamada T, Shinohara H, Akagi K, Shirakawa H (1986) J. Chem. Soc.; Chem. Commun.: 1315
620. Usuki A, Murase M, Kurauchi T (1987) Synth. Met. *18*: 705
621. Ozaki M, Peebles D, Weinberger BR, Heeger AJ, MacDiarmid AG (1980) J. Appl. Phys. *51*: 4253
622. Ebisawa F, Kurokowa T, Nara S (1983) J. Appl. Phys. *5*: 3255
623. Koezuka H, Tsumura A, Ando T (1987) Synth. Met. *18*: 699
624. Yoshino K, Kaneto K (1985) Mol. Cryst. Liq. Cryst. *121*: 247
625. Kaneto K, Agawa H, Yoshino K (1987) J. Appl. Phys. *61*: 1197
626. Gazard M (1986) In: Skotheim TA (ed) Handbook of conducting polymers. Dekker, New York, vol 1 chap 19
627. Noufi R (1983) J. Electrochem. Soc. *130*: 2126
628. Bull RA, Fan F-R, Bard AJ (1984) J. Electrochem. Soc. *131*: 687
629. Okabayashi K, Ikeda O, Tamura H (1983) J. Chem. Soc.: Chem. Commun.: 684
630. Tourillon G, Garnier F (1985) Mol. Cryst. Liq. Cryst. *121*: 305
631. Mizutani F, Iijima S, Tanabe Y, Tsuda K (1985) J. Chem. Soc.: Chem. Commun.: 1728
632. Miasik JJ, Hooper A, Tofield BC (1986) J. Chem. Soc.: Faraday Trans. I, *82*: 1117

Editor: H.-J. Cantow
Received July 19, 1988

Macromolecule Metal Complexes — Reactions and Molecular Recognition

Yoshimi Kurimura
Department of Chemistry, Ibaraki University, Mito, Ibaraki 310, Japan

In solution, macromolecule-metal complexes form microheterogenous regions occupied by the polymer-backbone where physicochemical properties differ from those of the bulk solution. Most of the significant reaction patterns of macromolecule-metal complexes are due to the characteristic nature of these microheterogenous regions. This article reviews recent studies of the reactions of macromolecule-metal complexes. For example, fundamental reactions such as complex formations, ligand substitution and electron transfer reactions, as well as interactions of metal complexes with biological substances, photochemical behavior, and molecular recognition are described.

Understanding of these fundamental reactions may help to design new functional materials such as nobel catalysts, compounds with biological activities, photo-conversion systems, semi-conducting or conducting materials, polymer modified electrodes, displays, sensors, and so on.

Advances in Polymer Science 90
© Springer-Verlag Berlin Heidelberg 1989

1 Introduction

Macromolecule-metal complexes, both synthetic and natural, have metal complex moieties on their polymer backbones. In solution, macromolecule-metal complexes form microheterogeneous regions occupied by the polymer-backbone where physico-chemical properties such as concentrations of reactive species, degree of solvation of the metal ions or metal complex moieties, structure and dielectric constant of the solvent, coordination number and structure of the metal complex, interactions of neighbouring groups and so on, are, in most cases, different from those of the bulk solution. Of course, the conformation of the polymer chain is also a very important factor for displaying cooperative effects or ability for molecular recognition. Most of the characteristic properties of macromolecule-metal complexes in solution is due to the formation of such microheterogeneous regions.

In the last decade, there has been a large number of reports on synthetic macro-molecule-metal complexes concerning their complexation, catalytic activities, redox reactions, adsorptions of gaseous molecules and metal ions, photochemical behavior, biochemical effects, modified electrodes, semiconductive and conductive materials, and so on.

In this review, however, only recent studies of the reactions of macromolecule-metal complexes will be reviewed. There have been some exellent reviews recently on macromolecule-metal complexes regarding syntheses, formation, characterization and catalytic activities [1], solar energy conversion [2], artificial oxygen carriers [3], and electrode processes [4]. Furthermore, the preprints of the 1st International Conference on Macromolecule-Metal Complexes "Tokyo Seminar on Macromolecule-Metal Complexes" were published in 1987 [5]. These reviews and the preprints give useful information about the recent development of the basic and applied chemistry of macromolecule-metal complexes.

2 Fundamental Reactions of Macromolecule-Metal Complexes in Solution

2.1 Formation of Macromolecule-Metal Complexes

Many interesting features have been observed in the complex formation between polymer ligands and metal ions. Coulombic, hydrophobic, and steric interactions are major factors governing the thermochemical and dynamic aspects of complex formation.

Complex formation of poly(amino acid)s and poly(amine)s with several metal ions has been studied by means of potentiometry, colorimetry, and spectrometry in aqueous solutions. In most cases, conformational change of the polymeric ligand is induced as the result of complex formation, leading to changes in reactivity.

Interaction of bivalent metal cations with poly(glutamic acid) has been extensively investigated [6-17], including the induction of α-helix formation by transition metal ions. Nevertheless, available data are not sufficient to explain the wide variety of metal-ion-induced conformational changes.

In a solution of fully neutralized poly(glutamic acid) in the absence of supporting electrolytes, α-helix formation is induced by the addition of bivalent transition metal ions [18]. The ability of α-helix induction follows the order: $Cu^{2+} > Cd^{2+} > Zn^{2+} > Ni^{2+} > Co^{2+}$. The induction of the α-helix formation by these metal ions is inhibited in the presence of large amounts of supporting electrolyte. On the other hand, induction of conformational change in the absence of supporting electrolyte does not occur with the addition of bivalent nontransition metal ions such as Ba^{2+}, Ca^{2+}, or Mg^{2+}. These metal ions are known to have little ability to coordinate to usual ligands. Distortion of circular dichroism (CD) spectra occurs as the formation of the α-helix proceeds, if the aggregation number (m) represented by Eq. (1) exceeds 100 [18]

$$m = \Delta R_\theta(C_M)/\Delta R_\theta(C_M = 0) \qquad (1)$$

where $\Delta R_\theta(C_M)$ and $\Delta R_\theta(C_M = 0)$ represent the difference of light intensities between solution and solvent at a scattering angle θ, and with the concentrations of M^{2+} equal to C_M and zero, respectively [18].

Investigation of the complex formation between copper (II) and a polymeric ligand derived from L-asparagine, poly-(N-methacryloyl-L-asparagine) (PMAsn) (1) as well as the corresponding low molecular weight ligand NIBAsn (2) showed that the latter ligand forms only a weak 1:1 complex through the carboxyl group [19].

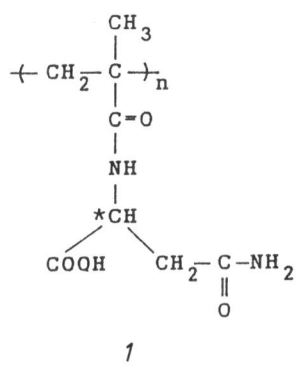

1

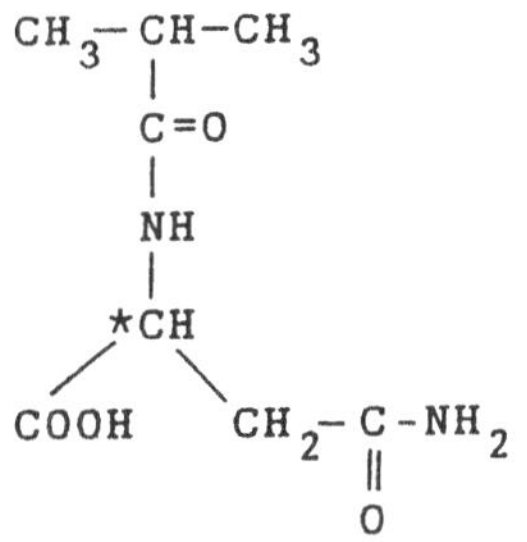

2

In contrast, PMAsn forms a series of different complex species depending on the pH. Some of these complexes involve the deprotonation of the amide group with formation of 1:1 species in one side chain (3), or 2:1 species between two side chains of the polymer. At high pH, complex formation with PMAsn and copper leads to a 2:1 complex (4) involving two carboxylate groups and, furthermore,

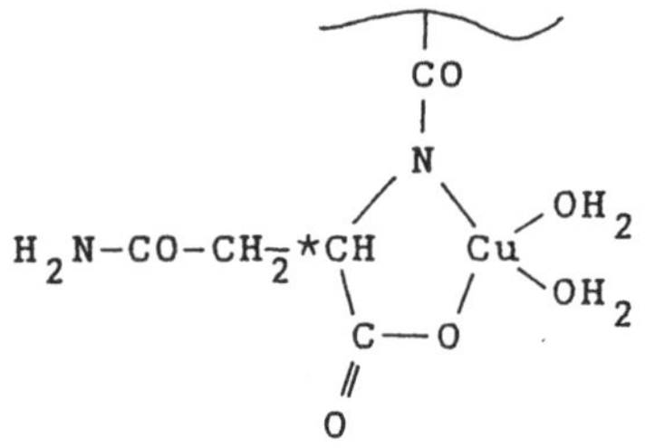

$$H_2N-CO-CH_2-\overset{*}{CH} \quad Cu \overset{OH_2}{\underset{OH_2}{\diagdown}}$$

with CO—N chelate and C—O (below), labeled

3

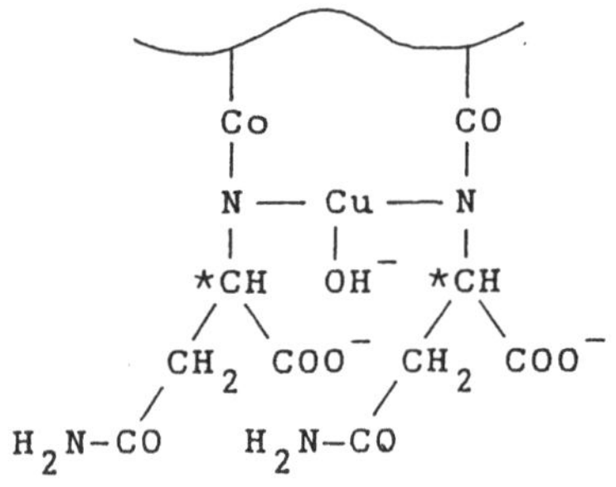

4

deprotonation of the amide group occurs. The difference between NIBAsn and PMAsn is due to the polymeric nature of the latter, where the vicinity of two side chains with carboxyl groups allows chelation. Another important effect is the high electrostatic field of the polyelectrolyte, which exerts a strong attraction on the divalent metal ion [19].

Binding behavior of transition metal ions with polyallylamine [20], poly(ε-N-methacryloyl-L-lysine) [21], branched poly(ethyleneimine)s [22], water soluble polymer-bound iminodiacetic acid analogue [23], and polyacrylamide gel [24] has been reported. In these works, the effect of the polymer backbone has been discussed in terms of interaction of metal ions with the polymer chains.

Chelating properties of linear and branched poly(ethyleneimine)s, LPEI (5) and

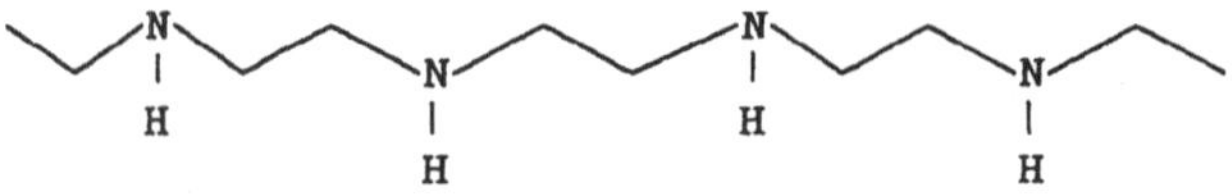

5

BPEI (*6*) respectively, have been studied to examine the chelating ability of poly(ethyleneimine)s having different microstructures. From the overall stability constant for Co^{2+}, Ni^{2+}, Cu^{2+}, Zn^{2+}, Cd^{2+}, and UO_2^{2+} ions, the chelating ability of LPEI was estimated to be ca 10 times less than that of BPEI for all but UO^{2+}. The branched structure gives a more favorable complexation than the linear structures [22].

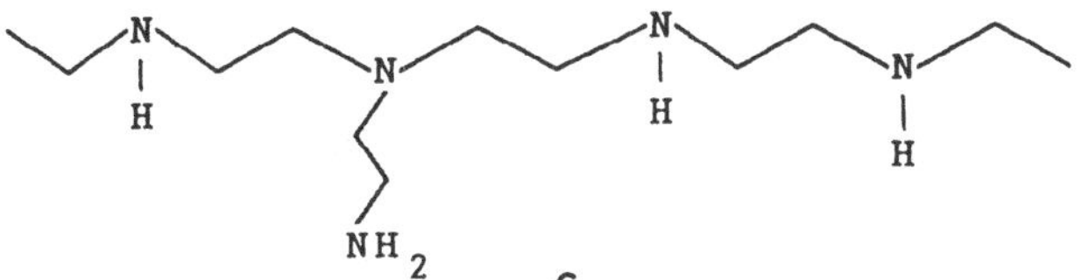

In some cases, crosslinking and/or aggregation of the synthetic polymer ligands proceeds as the result of their interaction with divalent metal ions [25]. Copolymers of poly(monoheptyl-co-diheptylitaconate) were modified by inserting up to thirteen tetraethylene-pentamine (tetrene) units per hundred monomer units in the chain. These modified copolymers are reacted with cobalt(II), and copper(II) chlorides. Polymer-metal complexes are formed by the interaction between the tetrene ligands and the transition metal ions leading to crosslinking of the samples. The formation of ion clusters was also detected, with a most common cluster dimension of 15 nm, although larger aggregates tended to form at higher metal concentrations. The thermal stability of the copolymer is improved after complex formation [25].

Complexation of poly(allylamine hydrochloride), $-(CH-(CH_2NH_3Cl)-CH_2)_n-$, with transition metal ions in aqueous solution appears to proceed in one step, in contrast to that of the corresponding low molecular weight metal complex. Interaction between polyallylamine and Cu^{2+}, Ni^{2+}, Co^{2+}, Zn^{2+}, and Mg^{2+} has been reported [21].

It was believed that cupric ions coordinate with poly(vinyl alcohol) (PVA) to give a polymer metal complex (*7*) with deprotonation of PVA [26, 27]. Reexamination of the

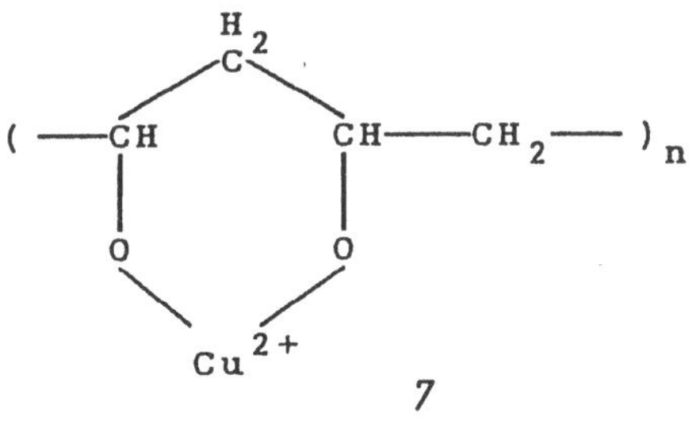

interaction between curpric ions and PVA by several spectroscopic techniques shows no usual complex as shown by (*7*) [28]. Curpic ions exist as the olated polynuclear complex of $Cu(OH)_2$ solubilized by PVA at pH > 6 and as ordinary hydrated ions at pH < 6. Solubilization occurs as the result of the inclusion of the $Cu(OH)_2$ aggregates into the PVA backbone due to hydrophobic interaction.

The interactions between cupric ions and poly(acrylic acid) (PAA) in aqueous solution have also been reexamined [29]. Cupric ions exist at pH < 3 as ordinary hydrated ions, and in the pH range 3–8, two mononuclear complexes with carboxyl groups of PAA, $Cu(OAc)^+$ and $Cu(OAc)_2$ (OAc = acetate anion), are formed in equilibrium with other complex species. Around pH = 4, two slightly different binuclear Cu(II) complexes are formed in high yield; at $[PAA]/[Cu^{2+}] < 10$, about 90% of the Cu^{2+} form such binuclear complexes. At pH > 6, $Cu(OH)_2$ is gradually formed and complex species other than $Cu(OH)_2$ finally disappear at pH > 9. The cupper hydroxide, $Cu(OH)_2$, does not deposit by hydrophobic interaction with PAA, as in the case with poly(vinyl alcohol).

The complexing ability of iminodiacetic acid type chelating agents, e.g. benzyl-aminediacetic acid (BDA) (8), was found to be enhanced considerably in aqueous solution by the use of a polymeric analogue such as (vinylbenzylamine diacetic acid)-co-(sodium styrenesulfonate) (PBDA) (9) [23]. The chelate formation constants for

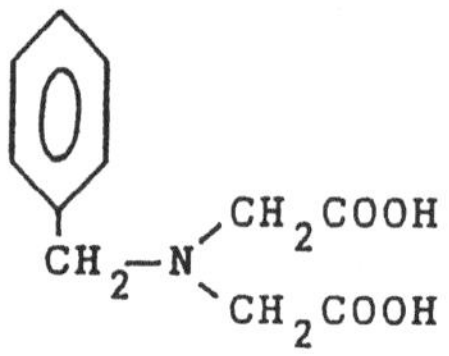

8

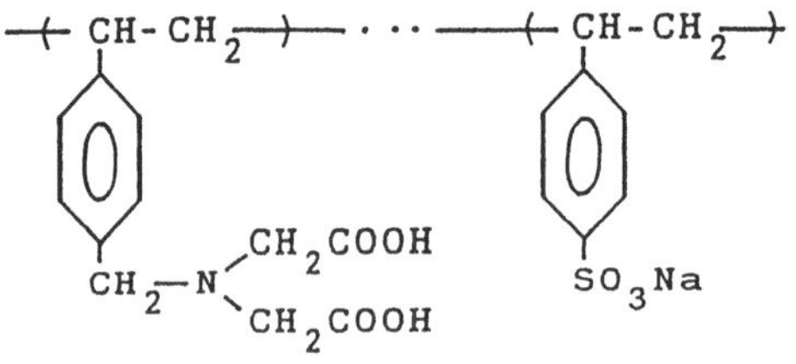

9

PBDA and BDA with some transition metal ions are shown in Table 1. For the polymer ligand, the complexation process appears to proceed in two steps: a) accumulation of M^{2+} ions into the PBDA domain (pre-equilibrium process) due to electrostatic interaction of the anionic polymer backbone of PBDA with M^{2+},

Table 1. Chelate formation constants of Co^{2+}, Ni^{2+} and Cu^{2+} with PBDA or BDA at ionic strength (I) = 0.1 and 25 °C[a]

Chelating agent	log K		
	Co^{2+}	Ni^{2+}	Cu^{2+}
PBDA	8.68	9.74	11.50
BDA	6.59	8.07	10.70

[a] 1.00×10^{-2} mol dm^{-3} M^{2+} and 1.00×10^{-2} mol dm^{-3} chelating agent

leading to polyion-metal ion associate and b) complexation of M^{2+} in the polymer domain with the chelating agent moieties (chelation process) (Eq. (2)).

$$PBDA + M^{2+} \xrightarrow[\text{process}]{\text{pre-equilibrium}} PBDA ---(M^{2+})_n$$
$$\text{(PBDA-metal associate)}$$

$$\xrightarrow[\text{process}]{\text{chelation}} M^{2+}(PBDA) \qquad (2)$$
$$\text{(metal chelate)}$$

With higher electrostatic potentials in the PBDA domain carrying their negatively charged groups closer to each other, the polymer chain can attract a larger number of divalent metal ions, leading to a high local concentration of M^{2+} ions in the domain. For PBDA, the higher formation constant is attributable to the existence of such a pre-equilibrium process. The "enhancement factor" (F), defined as $F = K_p/K_m$, where K_p and K_m are the metal complex formation constants for the polymer and low molecular weight chelating agents, respectively, is about 120 for Co^{2+}, 50 for Ni^{2+}, and 15 for Cu^{2+}, under the conditions employed. The results show that the enhancement factor decreases as the corresponding K_m increases (cf. Table 1). For metal complexes having larger K_m values, the effect of electrostatic accumulation of the metal ions in the polymer domains is smaller than for those having small K_m value, since relatively small amounts of the metal ions remain in the bulk of solution, as most metal ions are already combined with the chelating groups.

The rates of complex formation and ligand substitution reactions of the polymer-bound Co(III) complexes depend on the dynamic property of the polymer domains. Reports on the kinetics of complex formation and ligand substitution of macro-molecule-metal complexes are, however, relatively scarce. They include investigations on the complexation of poly-4-vinylpyridine with Ni^{2+} by the stopped conductance technique [30] and on a ligand substitution reaction of the polymer-bound cobalt(III) complexes [31].

The effect of polymeric ligands on the acid and base hydrolyses of cis-$[Co(en)_2PVPCl]^{2+}$ (PVP = poly-4-vinylpyridine, en = ethylenediamine) (10) and

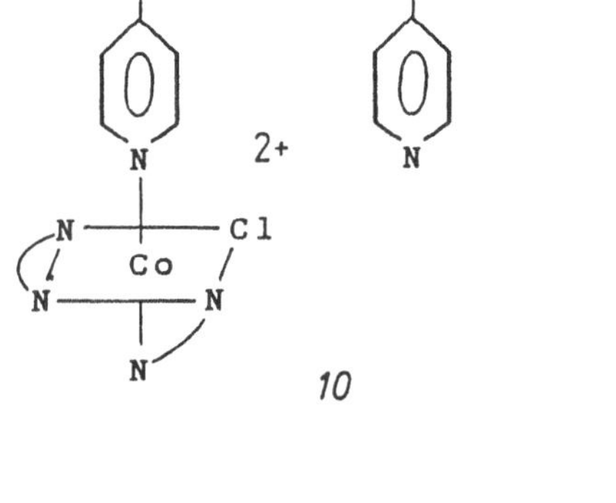

10

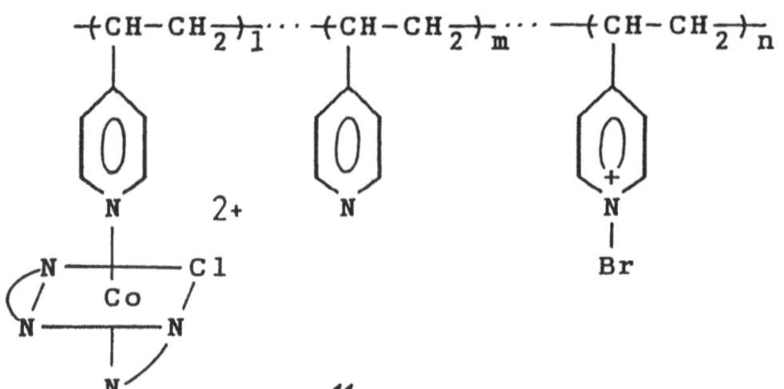

$$\begin{array}{ccc} +\text{CH}-\text{CH}_2\frac{1}{1} \cdots +\text{CH}-\text{CH}_2\frac{1}{m} \cdots & -\text{CH}-\text{CH}_2\frac{1}{n} \end{array}$$

11

cis-$[\text{Co(en)}_2\text{QPVPCl}]^{2+}$ (QPVP = partially quaternized poly-4-vinylpyridine by ethyl bromide) (11) has been investigated in aqueous solution [31]. For the hydrolysis reactions of the polymer and the corresponding low molecular weight complexes, cis-$[\text{Co(en)}_2\text{PyCl}]^{2+}$ (Py = pyridine), a linear dependence of the apparent rate constant (k_{app}) on the concentration of hydroxide ion, with a small intercept, is found, leading to the relationship

$$k_{app} = k_H + k_{OH}[\text{OH}^-] \tag{3}$$

where k_H and k_{OH} are the rate constants of the acid and base hydrolyses, respectively. The rate constants of the hydrolysis reactions are shown in Table 2. The smaller acid hydrolysis rate constant for the polymer complex is attributable to the presence of the bulky polymer chain attached to the Co(III) complex moieties: the polymer chain present in the neighbourhood of the complex moieties tends to break up the solvation shell of the complex, whereas the complex makes greater demands on solvation when passing through the process of charge separation. Thus, factors producing poorer overall solvation lead to lower hydrolysis rates [32]. On the other hand, the rate constants of the base hydrolyses of the polymer complexes are greater than that of the low molecular weight complex by a factor of about 3. This is ascribed to a greater local concentration of OH⁻ ions in the domains. Higher positive potential field in the polymer domain attracts large amounts of OH⁻ at lower concentrations of inert salt, leading to high local concentration of OH⁻ in the domain. In addition, the reactivity of the polymer complex is extremely

Table 2. Rate constants for the acid and base hydrolyses of cis-$[\text{Co(en)2LCl}]^{2+a}$

L	P_n^b	x^c	$k_H (s^{-1})^d$	$k_{OH} (\text{mol dm}^{-3} s^{-1})^e$
Py			2.4×10^{-6}	3.2×10^2
PVP	19	0.60		4.3×10^2
PVP	49	0.62		7.2×10^2
PVP	100	0.58	8.1×10^{-7}	8.7×10^2

[a] $[\text{Co(III)}] = 4.0 \times 10^{-3} \text{ mol dm}^{-3}$, [b] degree of polymerization of PVP, [c] degree of coordination of of Co(III) complex, [d] I = 0.02 and 40 °C, [e] I = 0.08 and 25 °C

sensitive to the concentration and kind of the inert salt. In the case of the low molecular weight system, the effect of the inert salt can be essentially explained by the primary salt effect. For the polymer system, the rate decreases drastically with increasing concentration of the counter anions: the decreasing tendency is more remarkable in sodium perchlorate solution than in the sodium chloride one. This is ascribed to enhancement of the shrinkage of the positively charged polymer backbone in the former solution. Higher reactivity of the polymer complex at low salt concentration is ascribed to a larger activation entropy, i.e. $\Delta H^{\pm}$ 96 kJ mol^{-1} and $\Delta S^{\pm} = 125$ JK^{-1} mol^{-1} for [Co(en)$_2$PyCl]$^{2+}$ and $\Delta H^{\pm} = 100$ kJ mol^{-1} and $\Delta S^{\pm} = 150$ JK^{-1} mol^{-1} for [Co(en)$_2$PVPCl]$^{2+}$, probably due to a large decrease in the degree of hydration of the Co(III) complex moieties in the course of the activation process.

Dynamic processes of complex formation of metal ions with poly-4-vinylpyridine (PVP) (Eqs. (4) and (5)) have been studied by means of the conductance stopped flow (CSF) and conductance pressure-jump (CPJ) technique [30].

$$Ni^{2+} + L \underset{k_b}{\overset{k_f}{\rightleftharpoons}} NiL^{2+} \qquad\qquad (4)$$

$$Ni^{2+} + HL^+ \underset{k_b'}{\overset{k_f'}{\rightleftharpoons}} NiL^{2+} + H^+ \qquad\qquad (5)$$

where L and HL$^+$ are the nonprotonated and protonated pyridine moieties on PVP, respectively. The observed values are $K_4 = 690$ M^{-1} (K = formation constant), $k_f = 4600$ M^{-1} s^{-1}, and $k_b = 6.7$ s^{-1} for Eq. (4) and $K_5 = 0.01$ M^{-1}, $k_f' = 10$ M^{-1} s^{-1}, and $k_b' = 1000$ M^{-1} s^{-1} for Eq. (5). The value of k_f is much lower than that for a diffusion-controlled process and the K, k_f, and k_b values of the Ni^{2+}/PVP system are quite similar to those of Ni^{2+} and a corresponding low molecular weight ligand (imidazole).

Poly(L-glutamic acid) (PLG) anion has been employed extensively as a model system because its identical, charged residues make it ideal for the interaction with counterions [33]. The poly(L-glutamate) helix can be used to test the line-of-charge model of Manning [34] or the charged-cylinder model [35] for ion condensation [33]. Europium(III) was used as the counterion of poly(L-glutamate) to investigate the interaction of PLG with the metal ion. The europium(III) ion differs from most other polyvalent ions because of its very narrow absorption band (~1 nm) and its fluorescence properties. Transitions from the F$_0$ ground state to a D$_0$ state (λ_{max} = 579 nm) or D$_2$ (λ_{max} = 465 nm) are environment sensitive.

The results of a fluorescence study with a solution of PLG-Eu(III) gave the following information [33]. First, intimate binding with loss of hydration water is possible only for the helical configuration, whereas free glutamate anions or anion residues in the random polymer do not induce a loss of Eu(III) water of hydration. Second, one half of the Eu(III) water of hydration is lost when the ion binds to the helical complex. Third, the transition from helix to coil at R = 8 (R = the number of glutamate residues per Eu(III) ion) is consistent with at least bidentate binding, which probably involves residues on adjacent helical groups. For larger R, the helices remain stable; for smaller R, charge-charge interactions induce the random configuration [33].

The binding between some metal and metal-complex ions and chiral polymers has been investigated to clarify the nature of the binding force, binding sites of the polymer chain, selective binding ability, recognition or resolution of optical isomers, and possibility of the metal or metal-complex ions as a chiral probe [36]. The selective nature of binding of polymer metal complexes has been utilized for selective cleavage of peptide bonds. $[Ru(bpy)_3]^{2-}$, $[Ruphen)_3]^{2+}$, and lanthanide ions were often used to accomplish some of these objectives. Fluorescence properties of lanthanide ions in aqueous solution are known to be strongly influenced by their immediate binding environment [37-41].

The fluorescence intensities and luminescence spectra of Tb(III) ($\lambda_{max} = 310$ nm), the hypersenitive band, are greately enhanced upon binding of these ions either to polyacrylate or to polysaccharides such as carboxymethylcellulose or heparin. In aqueous solution, Tb(III) ions are bound strongly to these molecules and some or all of the coordinated water molecules are expelled upon binding [40].

The fluorescence intensity and lifetime of the Tb(III)-polyelectrolyte complex have been measured [41]; the polyelectrolytes used were poly(acrylic acid), copolymers of maleic acid with ethylene (MA-E), with isobutene (MA-iBu), and 2,4,4-trimethylpent-1-ene (MA-3MPe). The fluorescence intensity decreases in the order MA-3MPe > MA-iBu > MA-E > PAA and the lifetime is in the reverse order.

The results show that the metal ion binding ability increases with increasing size of the alkyl group on the copolymer chains. The number of water molecules coordinated to the Tb(III) in polycarboxylates such as PAA and MA-E was found to be 3.5. On the other hand in the monomeric carboxylates such as propionate and 2,3-dimethylsuccinate this number was 5.8 to 6.0 [41].

The effect of conformation of poly(methacrylic acid) (PMA) and poly(acrylic acid) (PAA) on the photophysical and photochemical processes of $[Ru(bpy)_3]^{2+}$ has been reported [42]. A study of the pH induced phase transition of PMA and PAA suggests that, at high pH (pH 10), the Ru(II) is completely bound to the anionic stretched polymer, however, it is still in close contact with the aqueous phase. At intermediate pH, the lifetime and the fluorescence intensity of the Ru(II) show maxima at a pH of approx. 5. The probe also exhibits a blue spectral shift at this particular pH compared to other pH due to binding of the Ru(II) into the coiled or swollen polymer of PMA at pH 5. This binding is such that there is a restriction on the ligands of the metal complex probe so that, unlike in more mobile systems, complete relaxation of the excited state is not achieved. At pH 5, electrostatic side binding of cationic species exists inside the PMA coil which leads to an unusual photochemical process of $Ru(bpy)_3^{2+}$. The number of binding sites inside the PMA coil is far less than that on the surface of the PMA coil. Quenching of the excited state of $Ru(bpy)_3^{2+}$ by Cu^{2+} and Cr^{2+} is described by a model with a Poisson distribution of guest molecules among the PMA coils. Poly(methacrylic acid) shows a very sharp expansion transition at pH greater than 4, and at pH 5 the polymer is partly swollen and penetration of water into the polymer occurs. This enables polar cationic species to penetrate into the polymer and bind at internal sites thereby affecting the structure of the polymer coil. This type of behavior is minimized in polymers that do not show sharp transitions, such as PAA [42].

Tetrasodium tetrakis(p-sulfonatophenyl)-porphyrine (TTPS) associates strongly with poly(vinylpyrrolidone) (PVPRo) in aqueous solution. The equilibrium constant for the association of TPSP with PVPRo in pH 3 acetate at 25 °C is 1.4×10^7 m^{-1} s^{-1} for M = 48000 PVPRo. The lack of reactivity of TTPS bound to PVPRo indicates that the porphyrin microenvironment involves little contact with the bulk aqueous solution [43].

The effect of the polymer backbone on the intrinsic chemical reactivity of metal complexes has been studied in aqueous solution and in *Nafion* (perfluorocarbon sulfonic acid) film [44]. Using a model catalyst-substrate system, the independent kinetic effects of reaction site homogeneity, substrate diffusion into the polymer film, and changes on activation parameters have been addressed. The ligand substitution reaction (6), was chosen for this purpose (Py = pyridine and its derivatives).

$$Ru(NH_3)_5(H_2O)]^{2+} + Py = [Ru(NH_3)_5Py]^{2+} + H_2O \qquad (6)$$

The Ru(II) complex is electrochemically bound to Nafion polyelectrolyte films, and a variation of the electrode potential effects the conversion from the substitution-inert Ru(III) complex to the labile Ru(II) ion. The substitution of isonicotinamide for water in $[Ru(NH_3)_5(H_2O)]^{2+}$ was monitored by linear sweep voltammetry. The reaction activities for substitution by the various pyridines are generally 2 to 8 times faster than the rate observed in aqueous solutions containing identical concentrations of the ligand. In aqueous solution, the electrostatic attraction of $NC_5H_4COO^-$ yields a rate 10-fold higher than that for pyridine, yet the activity of $NC_5H_4COO^-$ in the polymer system is very low, and an upper limit for the rate constant is estimated as 0.029 l mol^{-1} s^{-1}. This indicates a change greater than 10^2 in the selectivity of Ru(II) for pyridine vs $NC_5H_4COO^-$ from aqueous media to the Nafion phase. The rates are independent of polymer thickness, indicating that the substitution rates are not limited by substrate diffusion into the polymer phase. A striking feature of these results is the contrast between the essentially constant reaction rate for substitution of the $[Ru(NH_3)_5(H_2O)]^{2+}$ ion with functionalized pyridine ligands in aqueous solution and the variation of reaction rate constants when the Ru(II) ion is immobilized in Nafion films. Examination of $\Delta H^{\neq}$ and $\Delta S^{\neq}$ for the substitution reactions of the three ligands indicates that the differences in intrinsic reactivity for the polymer and solution data are primarily due to an entropic effect.

2.2 Interactions of Metal Complexes with Biological Substances

Interactions of metal complexes with biological substances such as DNA and RNA are of major interest in the field of macromolecule-metal complexes.

Chiral ruthenium complexes, with luminescence characteristics indicative of binding modes, and stereoselectivities that may be tuned to the helix topology, may be useful molecular probes in solution for nucleic acid secondary structure [36].

The results of binding of the chiral metal complexes $[Ru(bpy)_3]Cl_2$, $[Ru(phen)_3]Cl_2$, and $[Ru(DIP)_3]Cl_2$ (DIP = 4,7-diphenyl- 1,10-phenanthroline) (12) to DNA showed that increasing luminescence is seen for $[Ru(phen)_3]^{2+}$ and $[Ru(DIP)_3]^{2+}$ with DNA addition, whereas no enhancement in luminescence is detectable for $[Ru(bpy)_3]^{2+}$ [45]. A biexponential decay in luminescence is found for $[Ru(phen)_3]^{2+}$ and $[Ru(DIP)_3]^{2+}$, with emission lifetimes of the complexes bound to DNA appearing 3 to 5 times

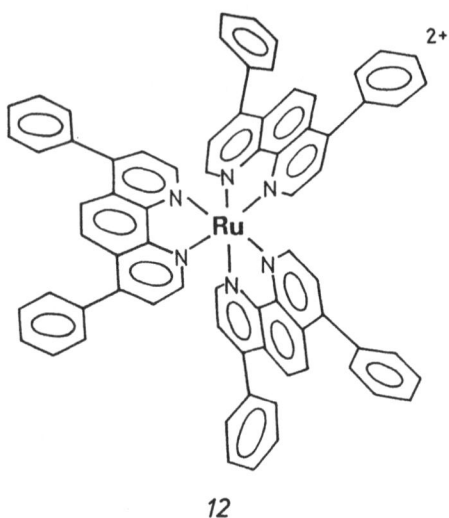

12

longer than those of the free complexes. Quenching of the luminescence by the ferro-cyanide anion further increases the ability to distinguish bound forms. *$[Ru(bpy)_3]^{2+}$ is quenched by $[Fe(CN)_6]^{4-}$ in the presence of DNA as efficiently as in its absence, indicating little or no binding. In contrast, biphasic Stern-Volmer plots are found for $[Ru(phen)_3]^{2+}$ and $[Ru(DIP)_3]^{2+}$, indicating extensive protection of their excited state complexes in the presence of DNA from ferrocyanide. Here emission quenching was found to be completely static as a result of counterion condensation at the DNA polyion. Emission polarization measurements revealed that the binding of $[Ru(phen)_3]^{2+}$ and $[Ru(DIP)_3]^{2+}$ to DNA is accompanied by a significant increase of the steady-state polarization. The results are interpreted in terms of two binding modes: electrostatic, which is essentially quenched by $[Fe(CN)_6]^{4-}$ and contributes no polarization in emission, and intercalative, which is protected from ferrocyanide quenching and, since rigidly bound, retains emission polarization. The distinction becomes more apparent for $[Ru(DIP)_3]^{2+}$ where significant enantiomeric selectivity is observed on binding to DNA. Thus Δ-$[Ru(DIP)_3]^{2+}$ binds to DNA both electro-statically and by intercalation: extensive curvature in the Stern-Volmer plots and increases in polarization are observed. The Λ-isomer, which gives strictly linear Stern-Volmer plots, binds only electrostatically. This chiral discrimination for inter-calative binding is explained in terms of the helical asymmetry of the right-handed DNA structure which is matched by the asymmetry of the Δ-isomer but precludes binding with the Λ-isomer.

Similar results were observed using $[Ru(phen)_3]^{2+}$ as a chiral probe for nucleic acids of different base compositions and structure and for A-type helices of DNA [46]. The structural characterization of two noncovalent binding modes of $[Ru(phen)_3]^{2+}$ to the DNA helix was achieved: one intercalatively bound mode showing a strong chiral preference for Δ-$[Ru(phen)_3]^{2+}$, and a surface-bound mode along the DNA groove, showing a weak preference for Λ-$[Ru(phen)_3]^{2+}$. Chiral discrimination for Δ-$[Ru(bpy)_3]^{2+}$ increase both with increasing Na^+ concentration and the percentage of guanine-cytosine (GC) pairs. The variations are attributed to changes in chiral preferences for intercalation. This variation may indicate local changes in DNA groove size, e.g. a compression along the helix axis direction with increasing ionic strength or increasing percent GC. Weak surface binding, having a preference for Δ-[Ru(phen)$_3$]$^{2+}$, was observed with double-strand RNA.

The ruthenium(II) complex [Ru(TMP)$_3$]$^{2+}$ (TMP = 3,4,7,8-tetramethyl-phenanthroline) also binds preferentially to the A-form helix, displays enanthiomeric discrimination in this binding, and upon irradiation with visible light cleaves A-form helices preferentially [36]. A plot of the ratio of bound metal per nucleotide vs the formal added ratio of metal per nucleotide indicates that the higest degree of binding is found with the double strand RNA; DNA-RNA hybrids show also cooperative binding to the ruthenium complex. In comparison, no binding to [poly[d(GC)] is detectable, and for a native, heterogeneous calf thymus DNA, at most a small level of binding is observed. Hence, [Ru(TMP)$_3$]$^{2+}$ is seen to bind cooperatively to the A-form polymer under conditions where little binding to B-DNA is detected. [Ru(phen)$_3$]$^{2+}$ and [Ru(bpy)$_3$]$^{2+}$, upon irradiation, serve as efficient singlet oxygen sensitizers, and photoactivated cleavage of A-DNA modified by singlet oxygen has been demonstrated for [Ru(bpy)$_3$]$^{2+}$ and [Ru(phen)$_3$]$^{2+}$. [Ru(TMP)$_3$]$^{2+}$ cleavages the DNA-RNA hybrid [poly(rC)-poly[d[$3H$]-G], whereas little cleavage is found for β-like poly[d[$3H$]-GC]. With Λ-[Ru(TMP)$_3$]$^{2+}$, twice the cleavage efficiency of the A-form polymer was observed in comparison with Δ-[Ru(TMP)$_3$]$^{2+}$.

2.3 Electron-Transfer Reactions

A large number of investigations of the mechanism of electron transfer reactions of macromolecule-metal complexes in biological systems has been reported. These investigations were concerned with not only natural metalloenzymes such as cytochromes, ferredoxin, blue coppers, oxygenase, peroxidase, catalase, hemoglobin, and ruberodoxin, but also modified metalloenzymes [47].

Electron-transfer in biological systems takes place through the mediation of a number of proteins, which contain a variety of active sites such as heme, Fe—S, Cu, and flavin. These active sites are protected from the solvent by a hydrophobic environment created by the peptide chain [48]. The redox potential of a biological redox couple in vivo lies, for the most part, between -0.5 and $+0.85$ V. The former and latter potentials correspond to the redox potentials of H_2O/H_2 and H_2O/O_2 respectively [49].

Characteristic features of the electron-transfer in biological systems are long-distance and directional electron transfer, and regulation of the rate of electron-transfer.

Direct evidence for long range electron-transfer in biological systems was first observed by Gray et al. [50, 51] and Isied et al. [48] using [Ru(NH$_3$)$_5$]$^{3+}$ substituted metallo protein. Histidine-83 of blue copper (azurin) was labeled with Ru(III)(NH$_3$)$_5$ [50]. Flash photolysis reduction of the His-83 bound Ru(III) followed by electron-transfer from the Ru(II) to Cu^{2+} was observed with a rate constant of 1.9 s^{-1}. The result shows that intramolecular long distance (approx. 1 nm) electron-transfer from the Ru(II) to the Cu^{2+} of the azurin takes place rapidly.

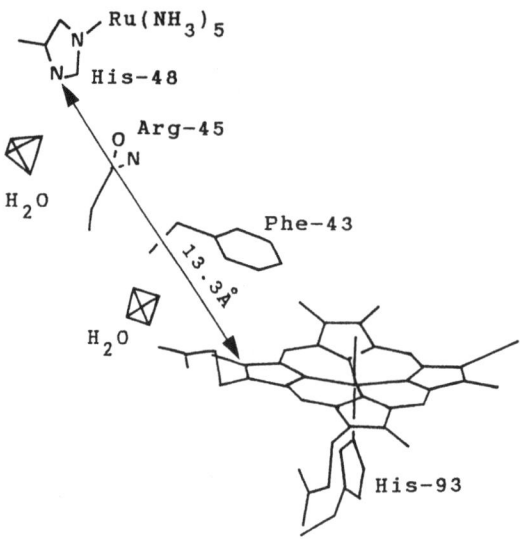

Fig. 1. View of selected parts of the molecular skeleton of $Ru(NH_3)_5$ modified(His-48) sperm whale myoglobin (from Ref. [52], modified)

Kinetics and thermodynamics of long-distance electron transfer in $Ru(NH_3)_5$ (His-48) myoglobin (Fig. 1, where selected parts of the molecular skeleton of sperm whale myoglobin with $Ru(NH_3)_5$ bonded imidazole of His-48 are illustrated) have also been determined [52]. The electron transfer observed is represented by

$$(NH_3)_5\ Ru(His-48)\ Mb(Fe^{3+}) \underset{k_{-1}}{\overset{k_1}{\rightleftarrows}} (NH_3)_5\ Ru(His\ 48)\ Mb(Fe^{2+}) \quad (7)$$

The rate constants (activation enthalpies) for the forward and backward reactions are $k_1 = 0.019 \pm 0.002\ s^{-1}$ ($\Delta H^{\ddagger} = 31 \pm 2\ kJ\ mol^{-1}$) and $k_{-1} = 0.041 \pm 0.003\ s^{-1}$ ($\Delta H^{\ddagger} = 81.5 \pm 0.4\ kJ\ mol^{-1}$) respectively. These results demonstrate that high-spin hemes are much less efficient in long distance electron transfer than low-spin analogue.

It was estimated that the enthalpic reorganization barrier for the heme in myoglobin is $84\ kJ\ mol^{-1}$, in contrast to the low-spin heme in the cytochrome c [53] analogue (29–33 $kJ\ mol^{-1}$). The reduction of metmyoglobin to deoxymyoglobin results in dissociation of axial water molecules from the iron atom. This change in ligation most likely accounts for the much larger reorganization barrier, because of the axial ligands (His and Met) in cytochrome c [54] are retained upon reduction of the iron center.

Intramolecular electron-transfers through peptides have also been observed by Isied and coworkers using $Ru(NH_3)_5$ modified cytochrome c [55]. Because of the kinetic inertness of both the ruthenium(II) and ruthenium(III), NMR and other physical techniques can be used to characterize the point of attachment of the ruthenium center. NMR and peptide mapping experiments showed that the ruthenium is bound to the His-33 site of cyt c (Fig. 2). The reduction potentials are $+0.26\ V$ for cyt c and $+0.07\ V$ for $[(NH_3)_5 Ru(His)]^{2+}$. Upon reduction of the Ru(III)-cyt c(III) derivative with 1 equiv. of electrons, any Ru(II)-cyt c(III) produced should undergo

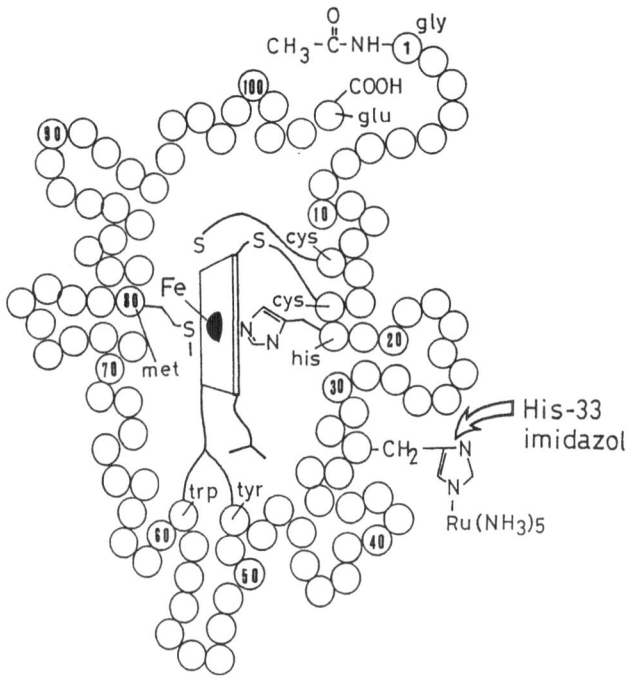

Fig. 2. Ruthenium(III) modified cytochrome c(His-33) derivative (from Eisenberg D, Kallai D, Sanson L, Copper A, Margolish E (1971) J. Biol. Chem. 246: 1511 and Ref. [55], modified)

intramolecular electron transfer to give Ru(III)-cyt c(II). The reduction was initiated by flash photolysis reduction of Ru(III) in the presence of carbon dioxide. The value of $82 \pm 20 \text{ s}^{-1}$ was obtained for the rate constant of the intramolecular electron-transfer from $Ru(NH_3)_5$ to cyt(III). The distance between the His-33 moiety and heme can be estimated to be 1.2 to 1.6 nm [55].

The rate constant for the reduction of cyt c(III) with cyt c peroxidase(II) was estimated to be $0.23 \pm 0.02 \text{ s}^{-1}$ [56]. The rate is independent of the initial (cyt c(III)/cyt c peroxidase) ratio, indicating that unimolecular electron-transfer takes place. When cyt c is replaced by porphyrin cyt c, a complex still forms with cyt c peroxidase. On radiolysis, using e_{ap}^- as the reducing agent, intracomplex electron-transfer occurs from the porphyrin cyt c anion radical to cyt c peroxidase(III) with $k = 150 \text{ s}^{-1}$. This large rate increase with increasing G_0 suggests that the barrier for intracomplex electron transfer is large. The reactivity of cyt c(III) of different origins follows the order: yeast > horse > tuna. The dependence of the rate of electron transfer on the primary structure of cyt c(III) is discussed [56].

Long range electron-transfer has also been demonstrated within the complex between zinc-substituted cytochrome c peroxidase and cyt c [59]. The kinetics of intramolecular electron-transfer from Ru(II) to Fe(III) in ruthenium modified cyt c has also been investigated [58].

Hemoglobin chains of one type (α or β) were substituted with closed shell zinc protoporphyrin (ZnP), and chains of the opposite type were oxidized to the

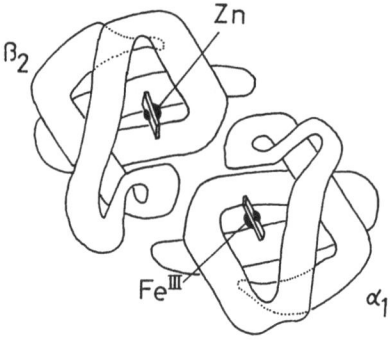

Fig. 3. [$\alpha(Fe^{III}H_2O)$; $\beta(Zn)$] hybrid

aquoferriheme ($Fe^{III}P$) state which is abbreviated as $\alpha'(Fe^{III}H_2O)\beta(Zn)$ (Fig. 3). Electron-transfer occurs within the $\alpha_1 - \beta_2$ complex between chromophores that are separated by two heme pocket walls and are at a metal-metal distance of 2.5 nm. Flash photoexcitation of ZnP to its triplet state (^3ZnP) initiates the primary process,

$$^3ZnP + Fe^{III}P \xrightarrow{k_t} (ZnP)^+ + Fe^{II}P(E - 0.8V) \ (k_t = 100 \pm 10 \ s^{-1}) \quad (8)$$

thereby forming an intermediate (ZnP^+) that returns to the ground state by back electron-transfer from $Fe^{II}P$ to the thermalized cation radical [Eq. (9)].

$$(ZnP)^+ + Fe^{II}P \xrightarrow{k_h} ZnP + Fe^{III}P(E - 1.0V) \quad (9)$$

A low temperature dependence of k_t indicates that the reaction involves electron and nuclear tunneling [59].

DNA mediated photoelectron transfer reactions have been demonstrated [60]. Binding to DNA assists the electron transfer between the metal-centered donor-acceptor pairs. The increase in rate in the presence of DNA illustrates that reactions at a macromolecular surface may be faster than those in bulk homogeneous phase. These systems can provide models for the diffusion of molecules bound on biological macromolecular surfaces, for protein diffusion along DNA helices, and in considering the effect of medium, orientation and diffusion on electron transfer on macromolecular surfaces.

Electron(hole) tunneling reactions are studied in a rigid polymer medium by following the reductive quenching of a series of $[*Ru(bpy)_3]^{2+}$ homologues by a series of aromatic amines. Tunnelling distances up to 1.2 nm (edge to edge) were observed [49]. These reactions are shown to be essentially temperature independent between 298 K and 359 K, but are significantly slower at 77 K.

The mechanism of the regulation of electron transfer in metalloproteins has been investigated [61] and two relevant examples have been discussed: in the first one the molecular mechanism controlling the electron transfer reactions is restricted to the immediate chemical environment of the metal center (azurin), while in the second one it involves a conformational transition of the whole quaternary structure of the enzyme. The power of the kinetic approach in detecting significant intermediates was emphasized [61]. The Cu metal complex site of azurin has a distorted tetrahedral

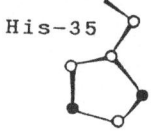

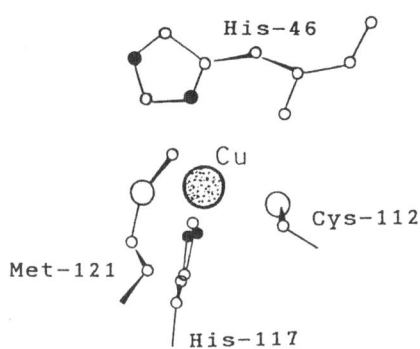

Fig. 4. The copper site in plastocyanin (from Adam ET, Canters GW, Hill HAO, Kitchen NA (1982) FEBS Letters 143: 287 and Ref. [61], modified)

array, very similar to plastcyanine, in which two sulfur atoms (Met 121, Cys 112) and two nitrogen atoms (His 46, His 117) of imidazole are included (Fig. 4) [61].

For plastcyanine [62], the nature of the ligands and the geometry of the site enhance the redox potential and are advantageous for an outer-sphere electron transfer mechanism [63] and with smaller molecular weight metal-protein [64]. The environment of the Cu center is hydrophobic, the metal beeing shielded from the solvent. Among the amino acid residues present in the immediate surroundings of the copper, His 35 is invariant [65], which is considered to play a crucial role in the switch mechanism from the active to the inactive form of reduced azurin: His 35 participates in a slow proton exchange process [66].

A model assuming full hydrogen bonding with groups of the protein of both the protonated and the deprotonated forms has been proposed [67], the conformational

$-\!(CH_2\!-\!CH)\!- \; -\!(CH_2\!-\!CH)\!- \; -\!(CH_2\!-\!CH)\!-$

13

transition between the two states being the rate limiting step. An alternative model [68] proposes a switch of an imidazole ring from one position, where it is accessible to water, to another inaccessible to water but suitable for hydrogen bonding [61, 68, 69].

Iron(III) chloro- and manganese(III) chloro-complexes of 5,10,15,20-tetrakis-(4-aminophenyl)21H,23H-porphin were covalently bound to a copolymer of ethenyl-benzene with (4-chloromethyl)ethenylbenzene (13) [70]. Electron transfer from the polymer-bound manganese(II) complex (13 for M = Mn(II)) to the polymer-bound iron(III) complex (13 for M = Fe(III)) has been investigated by means of stopped-flow spectrometry under pseudo-first order conditions in DMF. The rate constant for the polymer system is lower than that for a corresponding low molecular weight system. However, the activation energy for electron transfer between the polymer-bound metallo-porphyrins (92 kJ mol^{-1}) considerably lower than that corresponding to the reaction between the low molecular weight metalloporphrins (16.8 kJ mol^{-1}) [70].

Selective oxidations of optically active substrates by iron(II) complexes were observed in the oxidation of L-dopa and L-adrenaline by [Fe(tetpy)(OH)$_2$]$^+$ complex ions (14) anchored to poly(L-glutamate) (FeTL) or poly(D-glutamate) (FeTD) [74].

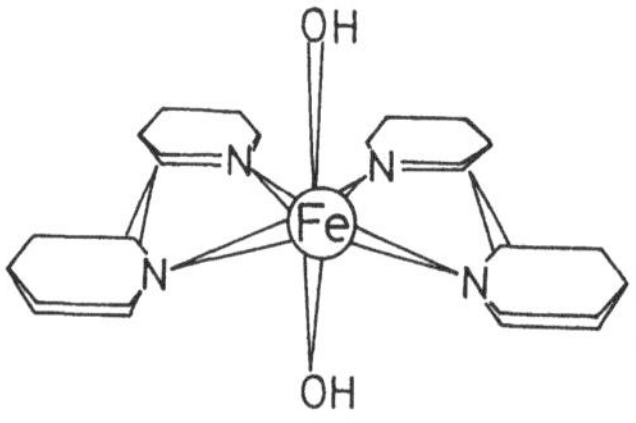

14

Electron transfer from the chiral catecholamines to iron(III) in the FeTL or FeTD system proceeds stereoselectively only when the polypeptide matrices are predominant-ly in the α-helical conformation and the accessibility of the active sites is, at least partially, hindered.

Intermolecular light-induced electron transfer to acceptors (quinones and nitro-benzenes) complexes in a β-cyclodextrin linked to a porphyrin has been studied and the dependence of electron transfer efficiency upon the reduction potential of the acceptor was examined [72].

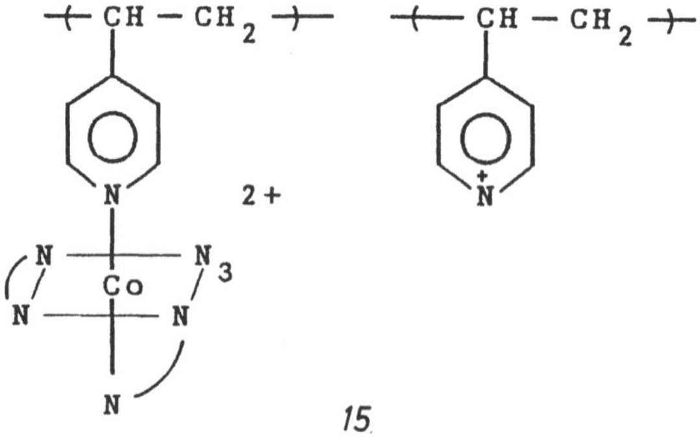

15

The rate of electron transfer of the polymer-bound Co(III) complex, cis-[Co(en)$_2$ PVP(N$_3$)]$^{2+}$ (PVP = poly-4-vinylpyridine, en = ethylenediamine) (15), with [Ru (NH$_3$)$_6$]$^{2+}$ is very sensitive to the type of dissolved innert anions at a given ionic strength [73]. In the domain, partial dehydration of the pendant Co(III) and, probably, [Ru(NH$_3$)$_6$]$^{2+}$ ions proceeds successively with an increase in the perchlorate ion concentration, leading to enhancement of the rate. The lower activation enthalpy observed in the perchlorate solution, relative to the chloride ion solution, is attributable to the rate enhancement in the former solution.

The theory of electron transfer in chemical and biological systems has been discussed by Marcus and many other workers [74–84]. Recently, Larson [81] has discussed the theory of electron transfer in protein and polymer-metal complex structures on the basis of a model first proposed by Marcus. In biological systems, electrons are mediated between redox centers over large distances (1.5 to 3.0 nm). Under non-adiabatic conditions, as the two energy surfaces have little interaction (Fig. 5), the electron transfer reaction does not occur. If there is weak interaction between the two surfaces, a$_1$ and a$_2$, the system tends to split into two continuous energy surfaces, A$_1$ and A$_2$, with a small gap Δ which corresponds to the electronic coupling matrix element. Under such conditions, electron transfer from reductant to oxidant may occur, with the probability ($\varkappa$) given by Eq. (10),

$$\varkappa = -\pi\Delta^2/2\hbar v(S_1 - S_2) \tag{10}$$

where v is a nuclear velocity, S$_1$ and S$_2$ are the slopes of the weakly interacting energy surfaces a$_1$ and a$_2$, respectively. This equation indicates that for $\Delta < 10^{-3}$ a.u. (approx. 2.4 kJ mol^{-1}) [81] the electron transfer is prohibited. For $\Delta > 10^{-3}$ a.u. the electron transfer becomes an adiabatic process, and hence can take place.

The Δ values of polymer metal complexes with different polymer bridges (16a–d), having redox couples M$_1$/M$_2$, where M$_1$/M$_2$ is Fe$^{2+/3+}$ or Cu$^{+/2+}$, at both terminals of the bridge, were estimated based on quantum mechanical treatment (Fig. 6). It was concluded that not only system (16a) but also system 16c may be adiabatic over

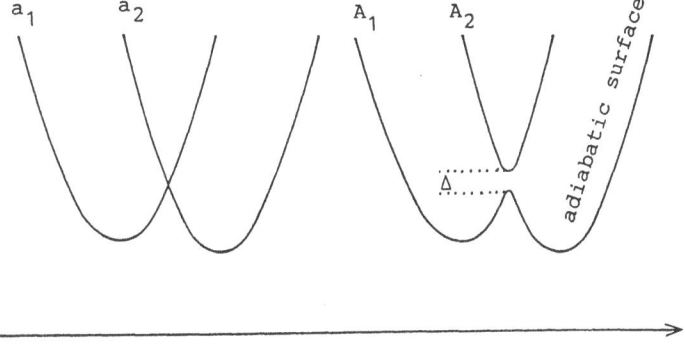

Fig. 5. Potential energy surfaces of reactants and products as a function of molecular configuration for an electron-transfer reaction

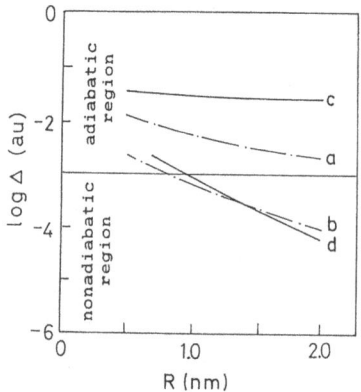

16 a

16 b

16 c

16 d

long distances provided the orbitals overlap in a suitable way. In the case of structures b and d, transfer becomes non-adiabatic at $R \simeq 1$ nm. Apparently the conformation is of decisive importance for electron transfer in this kind of bridge. The dynamical properties of the chain need to be known in order to decide whether the bridge is electron transferring or not [82].

A model has been given for electron tunneling in protein which allows the donor-acceptor interaction to be mediated by the covalent bonds between the amino acids and noncovalent contacts between amino acid chains [85].

Fig. 6. Relationship between Δ value and the distance R between two metal centers in structures *16a–d*

3 Catalytic Reactions

Important features of macromolecule-metal complexes as polymer bound catalysts are greater selectivity, recoverability, good facility for treatment, and, especially, appearence of novel catalytic activities.

Colloidal palladium or platinum supported on chelate resin beads were employed for the stereoselective hydrogenation of olefins [86]. Colloidal palladium supported on iminodiacetic acid type chelate resin beads was prepared by refluxing the palladium chloride and the chelate resin beads in methanol-water. Using the resin-supported colloidal palladium as a catalyst, cyclopentadiene is hydrogenated to cyclopentene with 97.1 % selectivity at 100 % conversion of cyclopentadiene under 1 atm of hydrogen in methanol at 30 °C. Finely dispersed metal particles ranging from 1 to 6 nm in diameter are the active species in the catalyst.

Colloidal platinum dispersions, prepated by photoreduction of tetrachloroplatinate(II) ion in the presence of a copolymer of N-vinyl-2-pyrrolidone and acrylamide, are treated with polyacrylamide gel having amino groups, resulting in stable immobilization of colloidal particles onto the gel. The immobilized catalysts exhibit high activities for the hydrogenation of olefins at 30 °C and 1 atm [87].

The Rh, Ru, and Pt ionomers of perfluoro- carbonsulfonic acid polymers have been formed and reduced to investigate the formation of metal particles within the ionic domains of these materials [88]. The particle size distributions peak in the 2.5 to 4.0 nm range. The reduced ionomers catalyze the CO oxidation with the activity sequence Ru > Rh > Pt. Diffusion limitations occur in the cases of the Rh and Ru, but not the Pt, ionomer catalysts.

Homogeneous, recoverable hydrogenation catalysts were prepared using functionalized ethylene oligomers as ligands [89]. Phosphine groups were introduced into the ethylene oligomers following anionic oligomerization of ethylene. The product poly-ethylenediphenylphosphine ligands were then exchanged with triphenylphosphine or ethylene ligands to prepare ethylene oligomer ligated rhodium(I) complexes. These Rh(I) complexes have the solubility of polyethylene and dissolve at 90 to 110 °C in hydrocarbon solvents, but quantitatively precipitate at 25 °C. Less than 0.1 % of the

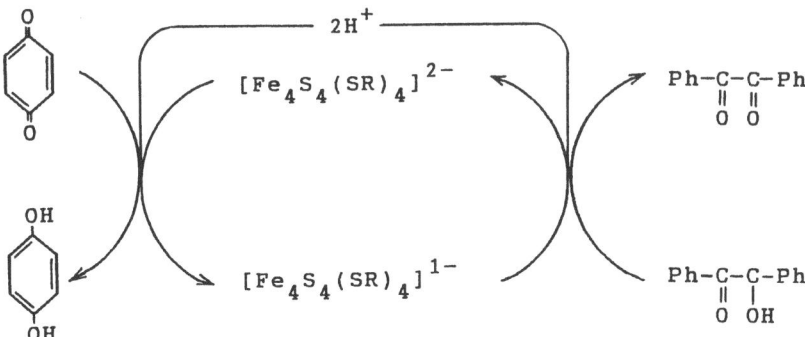

Fig. 7. Schematic representation of the catalytic oxidation of benzoin by 1,4-benzoquinone in the presence of $[Fe_4S_4(SR)_4]^-$

charged rhodium was lost in each dissolution and precipitation cycle. The rhodium(I) complexes so prepared were shown to have about 80% of the activity of tris(triphenylphosphine)rhodium chloride in hydrogenation of various alkenes including 1-octene, Δ-cholestene, cyclooctene, cyclododecene, styrene, α-methylstyrene.

Palladium complexes of chitin and chitosan have been shown to be active, selective, and stable catalysts for the hydrogenation of olefins, acrylic acid, nitrobenzene etc. at room temperature. Their catalytic activities can be controlled by changing the pH of the solution [90].

Studies of the catalytic or stoichiometric reactions of various 4-Fe-ferredoxin model complexes have been reported using the 3—/2— redox couple with a negative redox potential [91]. Relatively stable are the 2—/1— redox couple for [Fe(Z-cyst-Ile-Ala-OMe)$_4$]$^{2-}$ (Z = benzyloxycarbonyl) or [Fe$_4$S$_4$(tipbt)$_4$]$^{2-}$ (tripbt = 2,4,6-triisopropylbenzene-thiolato) in N,N-dimethylformamide. Catalytic oxidation of benzoin by 1,4-benzoquinone in the presence of various cyst-containing peptide complexes or bulky thiolato complexes in DMF has been examined. The postulated mechanism of the catalytic oxidation of benzoin in the presence of [Fe$_4$S$_2$(SR)$_4$]$^{2-}$ is illustrated in Fig. 7.

Oxovanadium(V) and oxomolybdenum(VI) were incorporated into crosslinked polystyrene resins functionalized with iminodiacetic acid or diethylenetriamine derivatives [92]. The polymer complexes were used as catalysts in the oxidation of olefins with t-butylhydroperoxide. Vanadium(V) complexes promote the epoxidation of allylic alcohols in a highly regioselective manner, e.g., 2,3-epoxide was obtained in 98% selectivity from ε-geraniol at 80 °C. The catalytic activity of the vanadium(V) complexes is generally higher than that of the molybdenium(VI) complexes in the oxidation of allylic alcohols, whereas an opposed trend holds for the epoxidation of cyclohexene.

A chiral ligand, analogous to 2,3-O-isopropylidene-2,3-dihydroxy-1,4-bis(dibenzophospholyl)-butane, DBP-DIOP, attached to linear and to crosslinked polymer supports (17) was obtained [94]. Exchange of Pt(II) (Pt(PhCN)$_2$Cl$_2$) onto the polymers (18), followed by the addition of stannous chloride gave polymersupported catalysts which were used for the asymmetric hydroformylation of a variety of olefins.

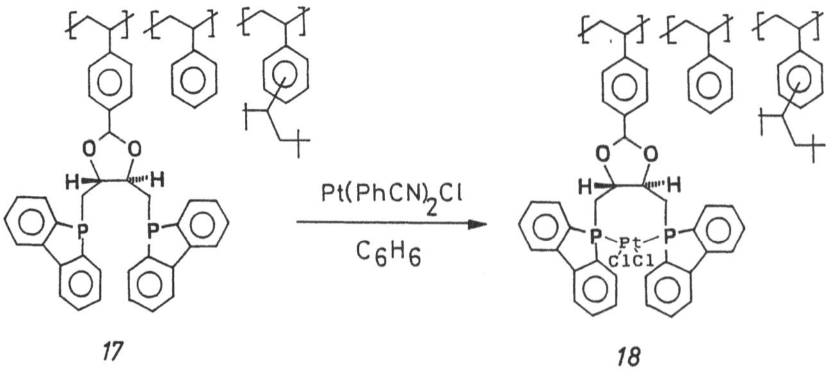

$$\text{(11)}$$

17   18

Hydroformylations utilizing the polymer-supported catalysts showed comparable rates and gave nearly the same optical yield as the homogeneous analogue. Recovering of the cross-linked polymer was achieved by simple filtration with slight loss in activity but no loss in selectivity [93].

Functionalized ethylene oligomer ligated rhodium(I) complexes were prepared and the hydrogenation of various alkenes including 1-octene, delta-2-cholestene, cyclooctene, cyclododecene, styrene, alpha-methylstyrene was studied [94].

A polymer-supported lipoamide-ferrous chelate system was used as catalyst for the reduction of diphenylacetylene to cis-stilbene with sodium borohydride: the dithiol-iron(II) (1:1) complex formed was suggested to be the active species. The chitosanlipoamide system has the highest activity among various insoluble polymers investigated [95, 96].

4 Photochemical Behavior

Luminescence properties of macromolecule-metal complexes have been well studied to investigate the characteristic behavior of the polymer domains to which the luminescent metal, or metal ion, or metal complex moieties are bound or associated.

Luminescence quenching of the zinc-substituted cytochrome c (Zn-cyt c) excited state by cytochrome b_5 (cyt b_5) has been investigated [97]. The most striking result of the quenching study is that the triplet emission decay rate of Zn-cyt c ($^3k = 10^2$ s^{-1}) is remarkably accelerated in Zn-cyt c/cyt b_5 ($^3k = 5 \times 10^5$ s^{-1}) by about a factor of 5×10^3. Several possible explanations for the dramatic increase in rate have been discussed.

The effect of temperature on the photoinduced electron transfer from [Ru(bpy)$_3$]$^{2+}$ to methyl viologen solubilized in cellophane has been investigated [98]. The first-order rate constant which depends exponentially on the distance between the reactants shows a non-Arrhenius type of behavior in the temperature interval from 77 to 294 K. This phenomenon, previously found to be of great importance in biological systems, is quantitatively interpreted in terms of a nonadiabatic multiphonon non-radiative process.

Quenching of the excited state of poly(acrylic acid-co vinylbipyridine)-pendant [Ru(bpy)$_3$]$^{2+}$ (19) by methyl viologen in alkaline aqueous medium has been studied and the effect of the polymer has been discussed [99]. Among the three different molecular weight samples used (MW 2100, 4400, 13300), the quenching of the lowest molecular weight polymer is explained to occur through a dynamic process, while the

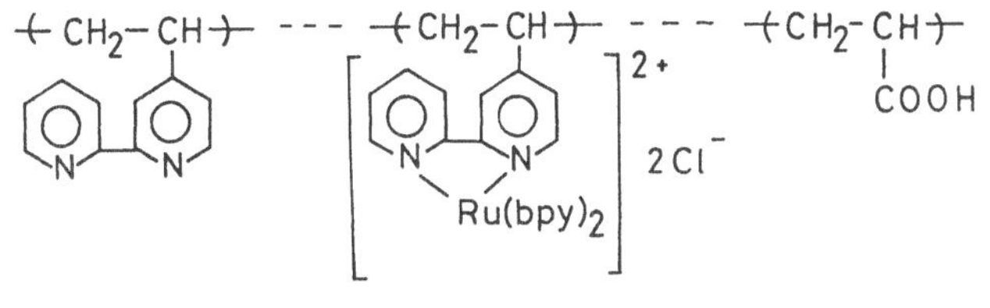

19

excited state of the complex with the higher molecular weight polymer (MW 4400) is quenched via both dynamic and static processes.

Analoguous $[Ru(bpy)_3]^{2+}$ complexes, consisting of polymers (20a) and (20b)

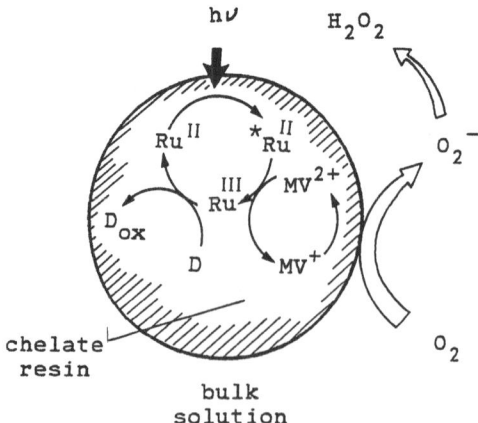

20a   R = H
20b   R = CH₃

were prepared and their luminescence behavior was investigated [100]. The absorption spectra and luminescence spectra of polymers (20a) and (20b) are essentially the same as that of $Ru(bpy)_3Cl_2$. The life time of 20a is very short (approx. 13 ns) in comparison to $Ru(bpy)_3Cl_2$ (660 ns), whereas the lifetime of polymer 20b (660 ns) is similar to that of $[Ru(bpy)_3]^{2+}$. The temperature dependence of the lifetime has been discussed in terms of Watt's model [101].

The $[Ru(bpy)_3]^{2+}$ photosensitized reduction of methyl viologen (MV^{2+}) proceeds rapidly in water-swollen iminodiacetic acid type chelate resin beads which adsorb both $[Ru(bpy)_3]^{2+}$ and MV^{2+} (RM resin). The reduction takes place with the aid of polymer-bound iminodiacetate as a donor [102]. Photosensitized formation of hydrogen peroxide occurs in an aqueous solution containing RM resin and oxygen molecules (Fig. 8) [102]. The some reaction also occurs using a polystyrene-coated filter paper, onto which both $[Ru(bpy)_3]^{2+}$ and MV^{2+} were adsorbed [103].

The quantum yield of the $[Ru(bpy)_3]^{2+}$ photosensitized reduction of Co(III)-Schiff base complex cation in aqueous solution is greately affected by the composition of the polymer-supported polyanionic donors such as vinylbenzylamine-N,N-diacetate-

Fig. 8. Schematic representation of the $[Ru(bpy)_3]^{2+}$ photosensitized reduction of dioxygen using water swollen chelate resin beads adsorbing $[Ru(bpy)_3]^{2+}$ and methyl viologen. Ru^{II}: $[Ru(bpy)_3]^{2+}$, *Ru: excited state of $[Ru(bpy)_3]^{2+}$, Ru^{III}: $[Ru(bpy)_3]^{3+}$, D: donor (benzylaminediacetate group)

co-styrene sulfonate (P-SS) (*21*) and vinylbenzylaminde-*N,N*-diacetate (P-VPRo) (*22*). The order of the quenching rate constants in the presence of a given concen-

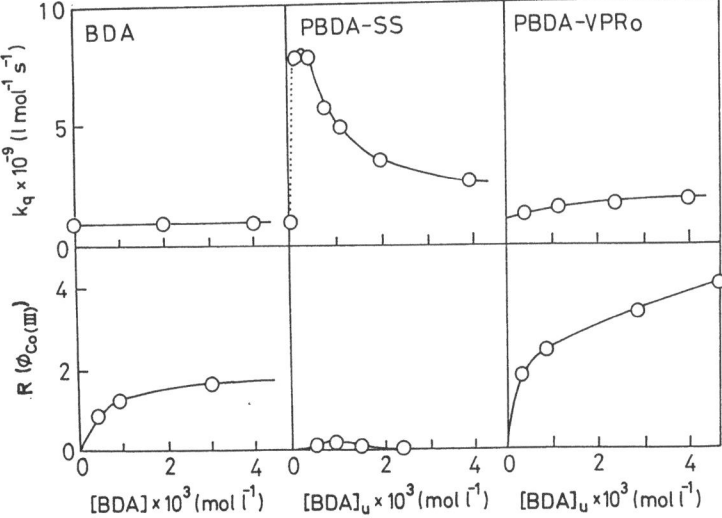

tration of the amine-diacetate analogues is P-SS ≫ P-VPRo > benzylamine-*N,N*-diacetate (BDA). On the other hand, the charge separation efficiency is in the order P-VPRo > BDA ≫ P-SS (Fig. 9) [104]. The results show that the rapid scavenging of chemically reactive oxidized species of the photosensitizer by the localized donor groups is of considerable significance in the design of the effective charge separation systems consisting of $[Ru(bpy)_3]^{2+}$/cationic acceptor/anionic polymer bound donor.

Fig. 9. Dependences of the quenching rate constant kg and the relative quantum yield, $R(\varnothing_{Co\text{III}})$, on the concentration of the various benzylaminediacetates

5 Adsorption of Gaseous Molecules and Molecular Recognition

Selective and reversible adsorption of gaseous molecules such as dioxygen, carbon monoxide, ethylene, acetylene, and dinitrogen have been performed by the use of suitable macromolecule-metal complexes. Selective adsorption of metal ions such as UO_2^- has also been studied using polymeric ligands.

Protoheme mono[N-[3-(imidazol-1-yl)]propyl]amide and mono[N_5-[3-(methylimid-azol-1-yl]pentyl]amide] were covalently bound with poly(1-vinyl-2-pyrrolidone) or dextran (abbreviated as PPH-1 and PPH-2, respectively). The polymer-bound hemes formed oxygen adducts with lifetimes of approx. 1 h for PPH-1 and a few minutes for PPH-2, in an aqueous ethylene glycol (1/1) solution cooled to −30 °C, whereas nonbound heme analogues do not form oxygen complexes under the same conditions [105]. The water-soluble but hydrophobic poly(vinylpyrrolidone) protects the heme-oxygen adduct from its proton-driven irreversible oxidation. The oxygen binding rate constant of PPH-1 is reduced by the surrounding polymer, and the oxygen binding affinity is close to that of hemoglobin [115].

The cobalt(III)-Schiff base complex, (N,N'-disalicyclidene-ethylenediamine)cobalt(II) (CoS), forms a five-coordinated complex with poly[(octyl methacrylate)-co-(4-vinylpyridine)] (POMPy) or poly[(octyl methacrylate)-co-(1-vinylimidazole)] (POMIm), which binds molecular oxygen rapidly and reversibly even in the solid state through the vacant sixth coordination site [106]. In membranes of the polymer-coordinated CoS complex, molecular oxygen transport is facilitated, because the CoS complex effectively acts as a fixed carrier for the oxygen transport in the membranes. Oxygen binding parameters determined spectroscopically are adequate to analyze the dual-mode transport model, by which the oxygen permeation behavior is quantitatively discussed.

Reversible oxygen binding has also been examined using poly(ethyleneimine)-cobalt complexes in aqueous solution. Cobalt(II) complexes of linear and branched poly(ethyleneimine) in aqueous solution are able to form with oxygen a μ-peroxo adduct as is evidenced from stoichiometry and spectral properties [107].

A novel reversible oxygen adsorbent system, consisting of a liposome and a heme-complex and serving as a physically stable oxygen carrier under physiological conditions, was prepared by Tsuchida et al [108]. The lipid-heme complex of 1-laurylimidazole was incorporated in polymerized 1-(p-vinylbenzoyl)-nonanoyl-2-o-octadecyl-3-phosphocoline (23). The diameter of the polymerized liposome/heme is approx. 35 nm. The oxygen adduct formation is rapid and reversible under physiological conditions even at high concentration. The rate constants for adsorption (k_{on}) and desorption (k_{off}) are $k_{on} = 1.3 \times 10^4 \, l \, mol^{-1} \, s^{-1}$ and $k_{off} = 0.42 \, s^{-1}$ at 25 °C and $p(O_2) = 40$ mmHg. These values are similar to those of blood.

Polystyrene-supported copper(I) chloride exhibits rapid and reversible adsorption of carbon monoxide [109]. For the adsorbent prepared by using toluene as solvent, the equilibrium molar ratio of adsorbed CO to the charged copper(I) chloride is 0.83 at 20 °C under 1 atm.

Desorption of CO is carried out at 7 mmHg, 20 °C, for 10 min. The equilibrium molar ratio of adsorbed CO to the charged copper(I) chloride in the second adsorption is 0.65. In the following adsorptions, the equilibrium molar ratios are virtually constant at 0.56. The adsorbing capacities of CO adsorbents prepared from aluminum

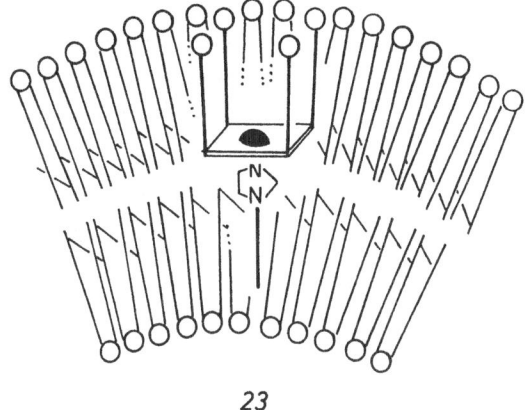

23

halides or copper(I) halides, with a macroreticular type crosslinked polystyrene, decrease as follows:

$$AlCl_3 \sim CuCl > AlBr_3 \gg AlI_3 \sim CuI$$

Chelate resin bead adsorbed Fe(II) can be used for reversible NO adsorption: the desired amount of $FeSO_4 \cdot 7\,H_2O$ was dissolved in water, and the resulting solution was added to the resin Diaion CR-10. After mixing for 24 h, the mixture was used as an adsorbent [110]. Although a Fe(II)-EDTA solution can adsorb NO rapidly even at low concentration of the Fe(II), the adsorption curve has a maximum at 25 min and then decreases with time probably due to oxidation of the Fe(II) ion with trace oxygen contained in the system. By use of ground resin beads, the adsorption of NO occurs rapidly and reaches equilibrium after 25 min. The adsorption rate was saturated at the molar ratio of one, indicating that the optimum molar ratio of charged Fe(II) to IDA is one.

Polymer-bound methylcyclopentadienyltricarbonylmanganese, $MeVCpMn(CO)_3$-co-R ($MeVCpMn(CO)_3$ = methylvinylcyclopentadienyltricarbonylmanganese, R = styrene or vinylpyrrolidone), $P\text{-}MeCpMn(CO)_3$, which is used as resin beads or membrane, reacts with acetylene to give stable acetylene complexes on irradiation with near ultraviolet- or visible light. [111, 112, 113)

$$\xrightarrow[C_2H_2]{h\nu} \qquad + \quad CO \qquad (12)$$

Reversible adsorption and desorption of the acetylene is possible under suitable conditions. However, using the corresponding low molecular weight manganese complexes such as $MeCpMn(CO)_3$ and $CpMn(CO)_3$ (Cp = cyclopentadienyl ring), the acetylene complex could hardly be prepared due to instability of the acetylene complex.

Direct synthesis of the polymer-bound dinitrogen complex, $P\text{-}MeCpMn(CO)_2$ (N_2), where the dinitrogen molecule is bound to the metal center by end-on type coordination, is also available.[114] On irradiation under nitrogen, $P\text{-}MeCpMn(CO)_3$ beads or film give the dinitrogen complex. The low molecular weight analog, $MeCpMn(CO)_2$ (N_2), could not be obtained by direct reaction; this dinitrogen complex can be prepared indirectly by Cu^{2+}/H_2O_2 oxidation of $CpMn(CO)_2$ (N_2H_2).

A template polymer complex, which incorporates N-benzyl-D-valine with almost 100% stereospecificity, has been synthesized by copolymerization of $\Delta\text{-}\beta_2\text{-}[Co\{(R,R')\text{-}$ N,N'-bis(4-(vinylbenzyloxy)salicylidene]-1,2-diaminocyclohexane}-(N-benzyl-D-valine)], styrene, and divinylbenzene, followed by dissociation of the coordinated amino acid[115].

New chelating polymers having high adsorption ability for uranium have been given much attention with respect to recovery and purification of uranium. The polymers having hydroxamic acid (*24*), amidoxime (*25*), dihydroxyphospino, or

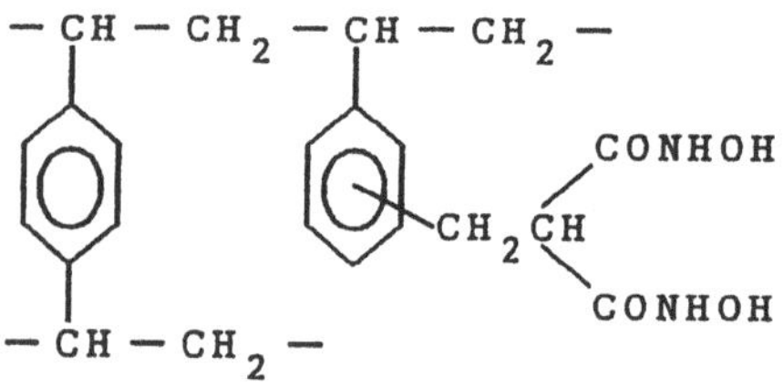

24

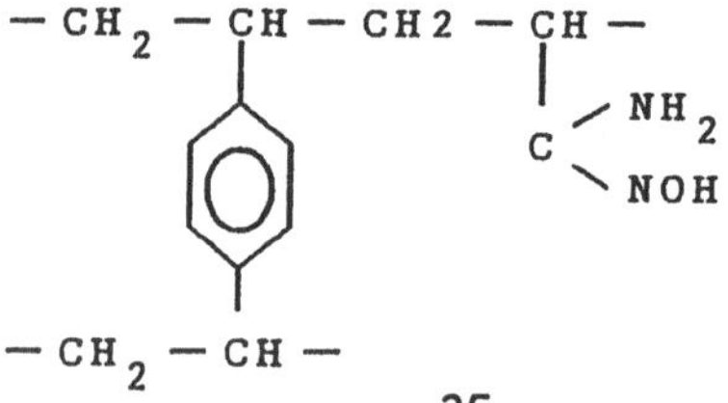

25

phosphono groups were prepared and the adsorptive properties for uranium were investigated.

The dihydroxamic acid polymer containing carboxylic acid groups as well as hydroxamic acid groups was used to examine the adsorptivity for uranium in seawater: the adsorptivity was 40 μg/g resin in 8 days[116].

Macroreticular chelating resins containing amidoxime groups (RNH) with various degrees of crosslinking were synthesized by using various amounts of ethylene glycol

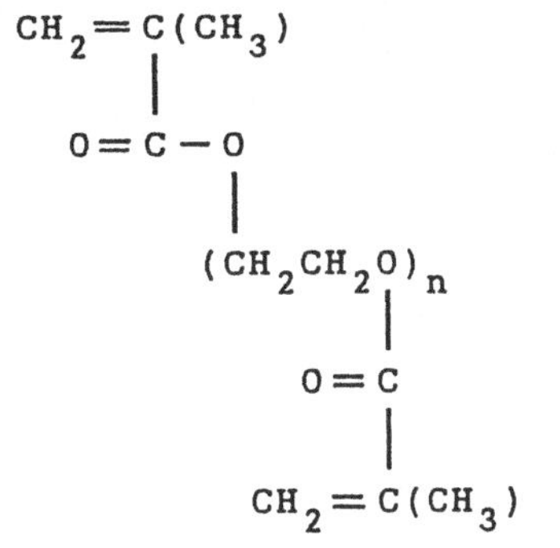

$$CH_2 = C(CH_3)$$
$$|$$
$$O = C - O$$
$$|$$
$$(CH_2CH_2O)_n$$
$$|$$
$$O = C$$
$$|$$
$$CH_2 = C(CH_3)$$

n	CR
1	1G
2	2G
3	3G
4	4G
9	9G

Crosslinking reagent (CR)

26

dimethycrylate, (1G)–(9G) (26), as crosslinking reagent [127]. The results suggested that the formation of not only favorable macropores but also of micropores is important for the effective recovery of uranium in seawater. Macrorecticular chelating resins containing amidoxime groups were synthesized from acrylonitrile (AN)-divinylbenzene(DVB), -alkyl acrylate, -alkyl methacrylate, or -vinylpyridine copolymer beads. A chelating resin containing amidoxime groups, and prepared from AN-DVB-methyl acrylate (MA) has the highest adsorption ability for uranium in seawater. After seawater was passed through the column packed with RNH-10-10 (10 mol% DVB and 10 mol% MA) at a space velocity of 180 h^{-1} (up flow) for 10 days, the amount of uranium adsorbed on the resin was 260 mg/kg [117,118]. Macroreticular chelating resins (27–30) containing dehydroxyphosphin and/or phosphono groups were prepared. The order of reactivity for uranium adsorption from seawater

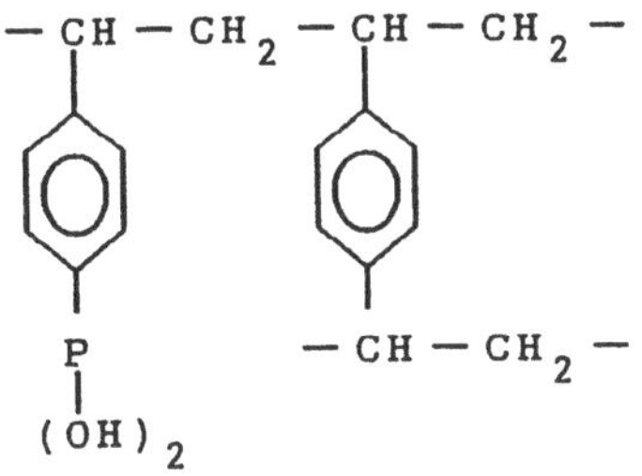

27

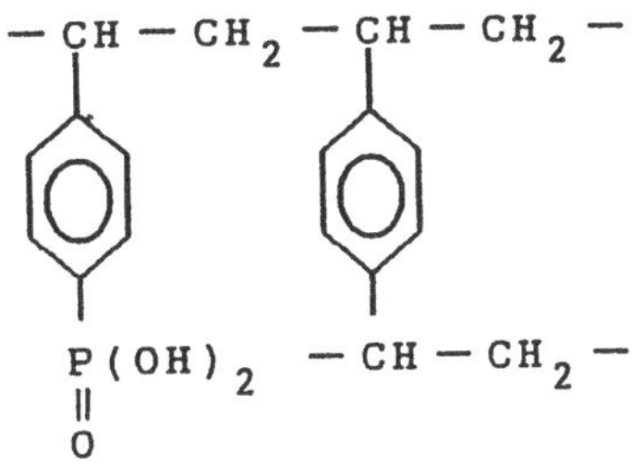

28

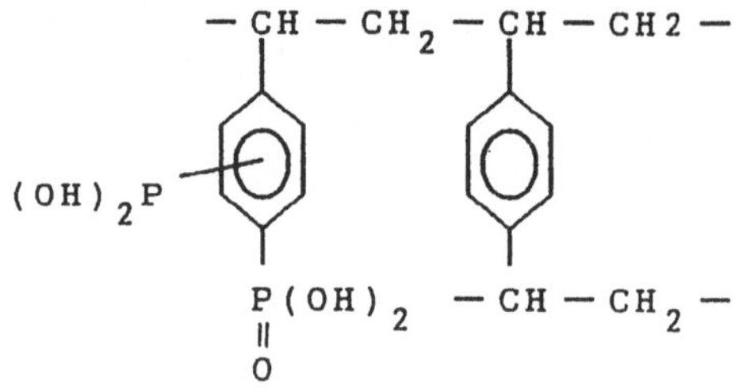

$$-CH-CH_2-CH-CH2-$$

(structure 29)

$$(OH)_2P$$

$$P(OH)_2 \quad -CH-CH_2-$$
$$\parallel$$
$$O$$

29

$$-CH-CH_2-CH-CH_2-$$

(structure 30)

$$(OH)_2P$$
$$\parallel$$
$$O$$

$$-CH_2 \qquad -CH-CH_2-$$
$$\mid$$
$$CH_2$$
$$\mid$$
$$P(OH)_2$$
$$\parallel$$
$$O$$

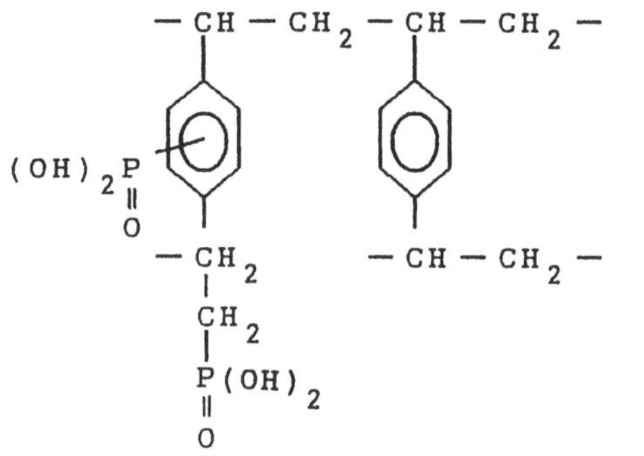

30

was (*30*) > (*29*) > (*28*) > (*27*). The resin (*29*) has a high physical stability and resistance against acid and alkali solution [119].

Certain natural restriction enzymes cleave double-helical DNA on opposite strands at or close to defined recognition sites, four to six base pairs in size [120–123]. Model compounds for such restricted enzymes are composed of multi-selective binding sites and Fe^{II} chelate sites which cleave the specific binding sites of base pairs of double helical DNA in the presence of molecular oxygen. To approach the design of sequence-specific molecules that read large sequences of DNA, polymer-bound metal complexes have been prepared by Dervan and his coworkers [123–129]. One example of such selective DNA cleaving molecule is bis(Fe^{II} EDTA-distamycine)-fumaramide (BEDF Fe^{II}) (Fig. 10): the dimer of tris-N-methylpyrrolecarboxamide connected by a flexible C-7 linker, which has the property of monomeric binding competitive with dimeric binding, was redesigned with a shorter rigid C-4 linker of favorable curvature affording a new molecule, BEDF, that has the property of exclusive dimeric binding which results in the recognition of eight to nine continuous base pairs of A-T rich double helical DNA [129]. A bleomycine model has also been demonstrated capable of recognition and selective cleavage of DNA molecules [130,131].

6 Future Studies of Macromolecule Metal Complexes

Techniques for the quantitative investigation of the dynamic processes involved, as well as of the physiochemical properties of macromolecular metal complexes, are

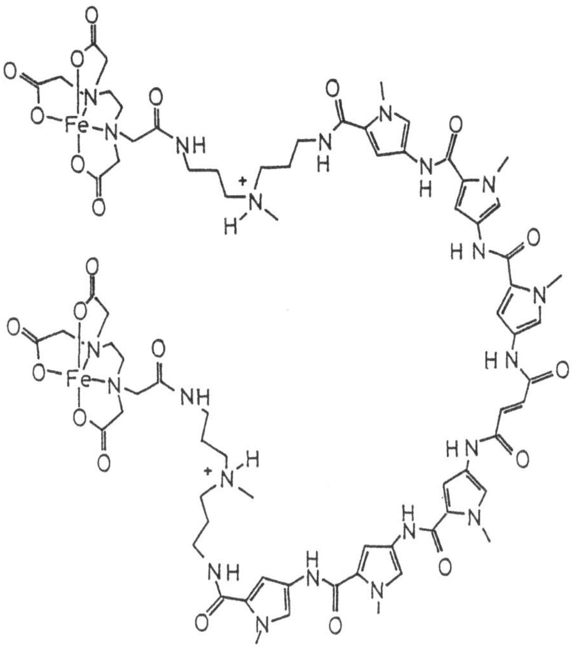

Fig. 10. Bis(Fe^{II}-EDTA-distamycin) fumaramide (BEDF-Fe^{II}) (from Ref. [125], modified)

being developed. Therefore, future intensive studies on these subjects will provide a wide variety of knowledge, in particular with regard to the microheterogeneous regions occupied by the polymer backbones, as well as the conformations and dynamic conformational changes of the macromolecule-metal complexes in solution. Moreover, an estimate of the concentration or activity, degree of solvation, and conformation of the reactive species, are of considerable significance for exact analyses of the reactions of macromolecule-metal complexes.

The chemistry of the macromolecule-metal complexes is a new field, connecting polymer and metal-complex chemistries. The number of papers in this field has increased rapidly during the last few years. Especially, electron-transfer, complex formation, photochemistry, electrode processes, catalysis, and molecular recognition seem to be major interests in the fundamental research in this field. The search for applications of macromolecule-metal complexes is another interesting subject, and further developments in this direction will bring about significant advantages with regard to the functionalization of conventional polymers and metal complexes.

7 Acknowledgement

The author would like to thank Professor S. Olivé and Dr. G. Henrici-Olivé for their kind suggestions.

8 References

1. Kaneko M, Tsuchida E (1981) J. Polym. Sci. Macromol. Rev. 16: 397
2. Kaneko M, Yamada A (1984) Advances in Polymer Science. 55: 1
3. Tsuchida E, Nishide H (1986) Topics in Current Chemistry, 132: 64
4. Kaneko M, Woehrle D (1988) Adv. in Polym. Chem. 84: 141
5. Preprints of the Tokyo Seminar on Macromolecule — Metal Complexes (1987)
6. Jacobson AL (1963) Biopolymers 1: 269
7. Jacobson AL (1964) Biopolymers 2: 207
8. Jacobson AL (1964) Biopolymers 2: 237
9. Takesada H, Yamazaki H, Wada A (1966) Biopolymers 4: 713
10. Kono N, Ikegami A (1966) Biopolymers 4: 823
11. Bianchi E, Bicchi A, Conis G, Ciferri A (1967) J. Macromol. Sci., A-1: 909
12. Yamaoka E, Masujima T (1979) Bull. Chem. Soc. Jpn. 52: 1286
13. Masujima T, Yamaoka K (1980) Biopolymers 19: 477
14. Imai N, Marinsky JA (1980) Macromolecules, 13: 275
15. Noji S, Yamaoka Y (1980) Macromolecules 13: 1114
16. Kurotu T, Kasagi M (1983) Polym. J. 15: 397
17. Mattice W, McCord R, Shippey PM (1979) Biopolymers 18: 723
18. Maeda H, Hiramatsu T, Ikeda S (1986) Bull. Chem. Soc. Jpn. 59: 587
19. Lekchiri A, Marcellet J, Marcellet M (1987) Macromolecules 20: 49
20. Gitzel J, Ohno H, Tsuchida E, Woehrle D (1986) Macromol. Chem. Rapid. Commun. 7: 397
21. Bekturov EA, Kudaibergenov SE, Zhaimina GM, Saltykov YP, Pel'menstein BA (1986) Macromol. Chem. Rapid Commun. 7: 339
22. Kobayashi S, Hiroishi K, Tokunoh M, Saegusa T (1987) Macromolecules 20: 1496
23. Kurimura Y, Takato K (1988) J. Chem. Soc., Faraday Trans. 1. 84: 841
24. Res GC, Shick S (1985) J. Phys. Chem. 89: 3598
25. Cowie JM, Wadi NMA (1985) Polymer, 26: 1566
26. Kuhn W, Toth I (1963) Z. Naturforsh 18A: 112
27. Kuhn W, Toth I (1963) J. Macromol. Chem. 60: 77
28. Yokoi H, Kuwata S, Iwaizumi M (1986) J. Am. Chem. Soc., 108: 3358
29. Yokoi H, Kuwata S, Iwaizumi M (1986) J. Am. Chem. Soc. 108: 3361
30. Okubo T, Hongo K, Enokida A (1984) J. Chem. Soc. Faraday Trans. I 80: 2087
31. Kurimura Y, Takagi Y, Saito M (1988) J. Chem. Soc. Faraday Trans. I 84: 1025
32. Basolo F, Pearson RG (1967) Mechanism of inorganic reactions in solution, 2nd ed, Weily, New York
33. Strazak M, Cohen M (1984) Biopolymers 23: 847
34. Manning G (1978) Q. Tev. Biophys. 2: 179
35. Record MT Jr, Anderson CF, Lohman TM (1987) Qurt. Rev. Biophys. 2: 103
36. Mai H-Y, Barton JK (1986) J. Am. Chem. Soc. 108: 7414
37. Gallagher EK (1964) J. Chem. Phys. 41: 3061
38. Yoshino N, Paoletti S, Kido J, Okamoto Y (1985) Macromolecules 18: 1515
39. Nagata I, Okamoto Y (1977) Macromolecules 77: 773
40. Crescenzi V, Britten HG, Yoshino N, Okamoto Y (1985) J. Polym. Sci., Phys. Ed. 23: 437
41. Okamoto Y, Kido J, Britain HG, Paoletti S (1987) Preprints of Tokyo Seminar on Macromolecule — Metal Complexes, Tokyo, p 67
42. Chu DY, Thomas JK (1985) J. Phys. Chem. 89: 4065
43. El Torki EM, Casano PJ, Reed WF, Schmehl RH (1987) J. Phys. Chem., 91: 3686
44. Lieber CM, Lewis NS (1985) J. Am. Chem. Soc. 107: 7190
45. Kumar CV, Barton JK, Turro NJ (1985) J. Am. Chem. Soc. 107: 5518
46. Barton JK, Goldberg JM, Kumar CV, Turro NJ (1986) J. Am. Chem. Soc. 108: 2081
47. Guarr T, McGuire ME, McLendon G (1985) J. Am. Chem. Soc. 107: 5014
48. Isied SS, Vassilian A (1984) J. Am. Chem. Soc. 106: 1726
49. Moore G, Williams RJP (1976) Coord. Chem. Rev. 18: 125
50. Margalit R, Kostic NM, Cho C-M, Blair DF, Chiang H-J, Pecht I, Shelton JR, Schroeder WA, Gray HB (1984) Proc. Natl. Acad. Sci. U.S.A. 81: 6554
51. Kostic NM, Margalit R, Che C-M, Gray HB (1983) J. Am. Chem. Soc. 105: 7765

52. Crutchley RJ, Ellis WR Jr, Gray HB (1985) J. Am. Chem. Soc. 107: 5002
53. Nocera DG, Winkler JR, Yocom KM, Bordingon E, Gray HB (1984) J. Am. Chem. Soc. 106: 5145
54. Takano T, Dikerson RE (1981) J. Mol. Biol. 153: 79
55. Isied SS, Worosila G, Atherton SJ (1982) J. Am. Chem. Soc. 104: 7659
56. Cheung E, Taykar K, Kornlatt JA, English AM, McLendon G, Miller JR (1986) Proc. Natl. Acad. USA 83: 1330
57. Ho PS, Sutoris D, Liang N, Margoliash E, Hoffman BM (1985) J. Am. Chem. Soc. 107: 1070
58. Norcera DN, Winkler JR, Yocom KM, Bordingnon E, Gray HB (1984) J. Am. Chem. Soc. 106: 5145
59. Peterson-Kennedy SE, McGourty JL, Hoffman BM (1984) J. Am. Chem. Soc. 106: 5010
60. Barton JK, Kumar CV, Turro NJ (1986) J. Am. Chem. Soc. 108: 6391
61. Brunori M, Colosimio A, Silvestrini MC (1983) Pure and Appl. Chem. 55: 1041
62. Colman PM, Freeman HC, Guss JM, Murata M, Norris VA, Ramshaw JAM, Venkatappa MP (1978) Nature 272: 319
63. Lappin GA, Segal MG, Weatherburn DC, Henderson RA, Sykes AG (1979) J. Am. Chem. Soc. 101: 2302
64. Wherland S, Pecht I (1978) Biochemistry, 17: 2585
65. Pydel L, Lundgren JO (1976) Nature 261: 344
66. Hill HAO, Smith BE (1979) J. Inorg. Biochem. 11: 79
67. Silvestrini MC, Brunori M, Wilson MT, Usmar VD (1981) J. Inorg. Biochem. 14: 327
68. Adman ET, Canters GW, Hill HAO, Kitchen NA (1982) FEBS Letters 143: 287
69. Powers L, Ching Y, Chance B, Muhoberac B (1982) Biophys. J. 37: 403A
70. Gitzel J, Ohno H, Tsuchida E, Woehrle D (1986) Polymer 27: 1781
71. Pispisa B, Palleschi A, Barteri M, Naradini S (1985) J. Phys. Chem. 89: 1767
72. Gonzalez MC, McIntosh AR, Bolton JR, Weedon AC: J. Chem. Soc., Chem. Commun. 1984: 1138
73. Kurimura Y, Kikuchi K, Tsuchida E (1987) Preprint of Tokyo Seminar on Macromolecule Metal Complexes, p 130
74. Marcus RA (1956) J. Chem. Phys. 24: 966
75. Marcus RA (1956) J. Chem. Phys. 24: 979
76. Marcus RA (1965) J. Chem. Phys. 43: 697
77. Marcus RA (1964) Ann. Rev. Phys. Chem. 15: 155
78. Hoffmann R (1963) J. Chem. Phys. 39: 1397
79. Sutin N (1968) Accounts Chem. Res. 1: 225
80. Hoppfield JJ (1974) Proc. Natl. Acad. Sci. USA, 71: 3640
81. Larson S (1981) J. Am. Chem. Soc. 103: 4043
82. Larson S (1983) J. Chem. Soc., Faraday Trans. 2. 79: 1375
83. Chance B, Devaut DC, Frauenfelder H, Marcus RA, Schriefer JR, Sutin N ed (1979), Tunneling in Biological systems, Academic Press Inc.
84. Moore GR, Williams RJP (1976) Coord. Chem. Rev. 18: 125
85. Beratan DN, Onuchic JN, Hopfield JJ (1987) J. Chem. Phys. 86: 4488
86. Boggess RK, Taylor LT (1987) J. Polym. Sci., Part A, Polym. Chem. Ed. 25: 685
87. Hirai H, Ohtaki M, Komiyama M: Chem. Lett. 1986: 269
88. Mattera VD Jr, Barnes DM, Chaudhuri SN, Risen WM Jr, Gonzalez RD (1986) J. Phys. Chem. 90: 4819
89. Bergbreiter DE, Chandran R (1987) J. Am. Chem. Soc. 109: 174
90. Huang HY, Dong TL, An Y, Wang XX, Jiang JJ (1987) Preprints of Tokyo Seminar on Macromolecule Metal Complexes, p 132
91. Ueyama N, Sugawara T, Kajiwara A, Nakamura A: J. Chem. Soc., Chem. Commun 1986: 434
92. Yokoyama T, Nishizawa M, Kimura T, Suzuki TM (1985) Bull. Chem. Soc. Jpn. 58: 3271
93. Parrinello G, Deschenaux R, Stille JK (1986) J. Org. Chem. 51: 4189
94. Bergbereiter DE, Chandran R (1987) J. Am. Chem. Soc., 109: 174
95. Nambu Y, Kijima M, Endo T (1987) Preprinte of Tokyo Seminar on Maclomolecule — Metal Complexes, p 193

96. Nambu Y, Kijima M, Endo T (1987) Macromolecules 20: 962
97. McLendon GL, Winkler JR, Nocera DG, Mauk MR, Mauk AG (1985) J. Am. Chem. Soc. 107: 739
98. Milosavljeciv BH, Thomas JK (1986) J. Am. Chem. Soc. 108: 2513
99. Kaneko M, Nakamura H (1987) Macromolecules 20: 2265
100. Sumi K, Furue M, Nozakura S (1984) J. Polym. Sci 22: 3779
101. Van Houten J, Watt RJ (1976) J. Am. Chem. Soc. 98: 4853
102. Kurimura Y, Nagashima M, Takato K, Tsuchida E, Kaneko M, Yamada A (1982) J. Phys. Chem. 86: 2432
103. Kurimura Y, Matsuo N, Kokuta E, Takagi Y, Usui Y: J. Chem. Res.(s), 1986: 238
104. Kurimura Y, Takato K, Takeda M, Ohtsuka N (1985) J. Phys. Chem. 89: 1023
105. Nishide H, Yuasa M, Hasegawa E, Tsuchida E (1987) Macromolecules 20: 1913
106. Tsuchida E, Nishide H, Ohyanagi M, Kawakami H (1987) Macromolecules 20: 1907
107. Nishide H, Yoshioka H, Wang SG, Tsuchida E (1985) Macromol. Chem. 186: 1513
108. Tsuchida E (1985) Macromol. Chem. Suppl. 12: 239
109. Hirai H, Hara S, Komiyama M (1986) Bull. Chem. Soc. Jpn. 59: 1051
110. Hirai H, Toshima N, Asanuma H: Chem. Lett., 1985: 655
111. Kurimura Y, Takagi Y, Tsuchida E (1987) J. Macromol. Sci.-Chem. A24: 419
112. Kurimura Y, Ohta F, Ohtsuka N, Tsuchida E: Chem. Lett. 1987: 1787
113. Kurimura Y, Ohta F, Gohda J, Nishide H, Tsuchida E (1982) Makromol. Chem. 183: 2889
114. Sellmann D (1971) Angew. Chem. Int. Ed. Engl. 10: 919
115. Fujii Y, Matsutani K, Kikuchi K: J. Chem. Soc. Chem. Commun., 1985: 415
116. Hirotsu H, Katoh S, Sugasaka K, Sakurai M, HIchimura K, Suda Y, Fujishima M, Abe Y (1986) J. Polym. Sci. Part A, 24: 1953
117. Egawa H, Nakagama M, Nonaka T, Yamamoto H, Uemura K (1987) J. Appl. Polym. Sci. 34: 1557
118. Egawa H, Nakayama M, Nonaka T, Sugiura E (1987) J. Appl. Polym. Sci. 33: 1992
119. Egawa H, Nonaka T, Iari H (1984) J. Appl. Polym. Sci. 29: 2045
120. Smith HO (1979) Science (Washington D.C.), 205: 455
121. Modrich P (1982) Crit. Rev. Biochem. 13: 288
122. Roberts RJ (1983) Nucleic Acids Res. 11: R135
123. Schultz DG, Dervan PB (1983) J. Am. Chem. Soc. 105: 7748
124. Schultz PG, Taylor JS, Dervan PB (1982) J. Am. Chem. Soc. 104: 6861
125. Schultz PG, Dervan PB (1983) Proc. Natl. Acad. Sci. USA 80: 6834
126. Taylor JS, Schultz PG, Dervan PB (1984) Tetrahedron 40: 457
127. Schultz PG, Dervan PB (1984) J. Biomol. Struc. Dyn. 1: 1133
128. Youngquist RS, Dervan PB (1985) Proc. Natl. Acad. Sci. USA 82: 2565
129. Youngquist RS, Dervan PB (1985) J. Am. Chem. Soc. 107: 5528
130. Falab S (1967) Angew. Chem. 79: 500
131. Kilkuskie RE, Suguna H, Yellin B, Murugesan N, Hecht SM (1985) J. Am. Chem. Soc. 107: 260

Editors: Dres. Olivé
Received Juli 27, 1988

Molecular "Recognition" in Interpolymer Interactions and Matrix Polymerization

I. M. Papisov and A. A. Litmanovich

Moscow Automobile And Road Construction Institute,
64, Leningradsky prospect, Moscow, 125829, USSR

Results of theoretical and experimental investigations into selectivity in interpolymer reactions (molecular recognition) and the role played by this selectivity in the reactions of matrix synthesis of polymers have been systematized and generalized in this review.

This fundamental property of interpolymer interactions — molecular recognition — results from the exponential dependence of polycomplex stability on the lengths of interacting chains. Recognition is manifested in the selectivity of interpolymer interactions with regard to the length and the structure of the chains which, under certain conditions, allow us to fractionate efficiently non-uniform polymers as well as polymer mixtures. Inversion of molecular recognition is possible when different polymer partners are recognized under different conditions.

The mechanism of molecular recognition reactions (kinetics and mechanism of substitution and exchange reactions with participation of polycomplexes and free chains) is considered.

As a necessary matrix-process step of polymer synthesis, molecular recognition of macromolecular-matrices by the growing daughter chains has an influence on the kinetics of these processes and on the structure of the daughter chains. In systems of two or more matrices with different chemical structures, molecular recognition predetermines the growth of daughter chains only on one (the strongest) matrix.

Systems are considered in which, under certain conditions, matrix regeneration is possible in the process of matrix synthesis (i.e. freeing of the matrix from the polycomplex being formed) and multiple use of a single matrix for controlling the growth of the daughter chains.

Advances in Polymer Science 90
© Springer-Verlag Berlin Heidelberg 1988

List of Basic Abbreviations

AA   — acrylic acid
DP   — degree of polymerization
MA   — methacrylic acid
MWD   — molecular weight distribution
p.c.   — polycomplex
PAA   — polyacrylic acid
PAAm — polyacrylamide
PEI   — polyethylene imin
PEO   — polyethylene oxide
PFU   — copolymer of urea and formaldehyde
PMA   — polymethacrylic acid
PPh   — polyphosphate
PVA   — polyvinyl alcohol
P4VP   — poly-4-vinylpyrridine
P2VP   — poly-2-vinylpyrridine
PVP-q — partly or completely quaternised with ethylbromide P4VP;
    q — degree of quaternization, mole %
PVPD — polyvinylpirrolidone
PVSA — polyvinylsulfoacid
4-VP   — 4-vinylpyrridine

1 Introduction

Brought to perfection, the selectivity of interaction between macromolecules, i.e., their ability to "recognize" a certain partner, makes the basis for forming composite molecular structures, which play a decisive role in living organisms. This ability may be assumed to have developed in the evolutionary process and, hence, it might be characteristic not only of biopolymers and their synthetic analogues but of simple macromolecules as well.

The molecular "recognition" ability of simple synthetic macromolecules is of interest from the viewpoint of forming both highly efficient polymer reagents and new compsite materials — interpolymer interaction products. The principle of molecular "recongition" plays an important part in fractionation of polymers and highly dispersible systems, separation of composite mixtures of high molecular wight compounds — both of synthetic and of biologic origin, in precise "addressing" of physiologically active macromolecules to certain molecular structures in living organisms, in flocculation and dehydration of suspensions; molecular "recognition" in the processes of polymer matrix synthesis exerts influence on the structure (and hence on the properties) of both the forming macromolecules and the synthesis products — interpolymer complexes (polycomplexes) of these macromolecules with macromolecules of the matrix.

Results of experimental and theoretical investigations of interpolymer interactions, matrix polymerization and properties of interpolymer complexes have been summarized in various reviews [1-5] and monographs [6-7]. These were to some extent stimulated by successes in studying matrix synthesis in biological systems and peculiarities of intermolecular interactions with participation of biopolymers and their synthetic models. That many principles of intermolecular interactions, regardless of the nature of interaction macromolecules, are found to be common causes little surprise today, these being connected simply with the polymer nature of reagents. Historically many principles found while investigating biopolymers and their models (for instance, polynucleotides), were considered newly "discovered" when studying the simplest polymer systems. However, the significance of such secondary "discoveries" is of principal importance, emphasizing the fundamental character of a whole number of peculiarities pertaining to macromolecular system — primarily their ability for molecular "recognition" and selforganisation. According to Oparin and his colleagues [8-10], knowledge of the principles and mechanism of molecular "recognition" at the level of simple polymer systems can be of key significance for understanding the causes which determined the speed of the prebiological evolution of macromolecules.

Sufficient experimental and theoretical data have been already obtained permitting us to consider the molecular "recognition" ability, a fundamental property of macromolecular systems; this property manifests itself under interactions where simple synthetic macromolecules (homo- and copolymers) participate, if they result in the formation of a polycomplex, i.e., a compound of the following type

$$\sim A - A - A - A \sim$$
$$\vdots \quad \vdots \quad \vdots \quad \vdots$$
$$\sim B - B - B - B \sim$$

Scheme 1

stabilised by a cooperative system of intermolecular interactions between two macromolecules (containing, accordingly, fragments A and B in the main chanis or side groups). The nature of A ... B type interactions may be different; they may be hydrogen bonds, electrostatic interactions, etc. Type (1) structures result from interactions between previously synthesised macromolecules or during matrix polymerisation and polycondensation processes.

2 Molecular "Recognition" in Interpolymer Interactions

In a limited case, molecular "recognition" in dilute solutions of polymers may be schematically represented as

$$P + \sum_{i=1}^{n} P_i \rightarrow \text{p.c.} (P \cdot P_K) + \sum_{i=1}^{k-1} P_i + \sum_{i=k+1}^{n} P_i \qquad \text{Scheme 2}$$

where P is a macromolecule which, under given conditions, is capable of binding in polycomplex p.c. $(P.P_i)$ with any of the macromolecules $P_1, P_2, ... P_n$ present in the system:

$$P + P_i \rightarrow \text{p.c.} (P.P_i) \qquad \text{Scheme 2a}$$

which differ from each other in chemical structure, DP or both. Obviously, macromolecule P selects its optimum partner P_k only if two conditions are fulfilled. One is thermodynamic: p.c. $(P.P_k)$ must be the most stable one of them all, i.e., formation of this complex must correspond to the minimum free energy of the system. The second condition is kinetic: it is obvious that owing to random encounters, macromolecule P binds with any chain P_i, hence the choice of the optimum partner P_k must take place in a process of multiple "trials and errors" within a reasonable period of time, i.e., intermacromolecular substitution reaction:

$$\text{p.c.} (P.P_i) + P_j \rightarrow \text{p.c.} (P.P_j) + P_i \qquad \text{Scheme 3}$$

where $i \neq j$ runs at a sufficiently high rate. This condition is trivial, its fulfilment, however, is not self-evident as polycomplexes are quite stable compounds which reversibly dissociate only with small chain lengths of one of the components or in sufficiently narrow intervals of pH medium, ionic strength, etc [1,6]. That is why finding the substitution reactions has been of principal significance. Such reactions were first observed in works [11,12] where it was established that in diluted aqueous solutions PVPD substitutes PEO and PVA in their complexes with polyacids and PMA substitutes PAA in its polycomplex with PEO. Accordingly, polyacids "recognized" the "strongest" partner — PVPD in mixtures of PVPD with PEO and PVPD with PVA, while PEO in mixtures of PAA with PMA substituting for the latter. It is obvious from Table 1 that quite a number of reactions of "recognition" and substitution in many diverse systems have already been studied up to the present.

Some of the above examples show interpolymer interactions to be highly selective — "recognition" may require, however, insignificant differences in the copolymer composition of even differences in the structure of stereoisomers (see examples 10–12 in Table 1).

In its general form, the problem of molecular "recognition" boils down to finding the relative probability of binding macromolecule P with macromolecules $P_1, P_2, \ldots P_n$ present in the mixture if each of them, taken separately, is capable of interacting with P. If all macromolecules P_i are brought together under the term "nonuniform polymer $\tilde{P}$" and a continuous or discrete distribution function $\tilde{P}$ is introduced by nonuniformity parameters $w(\{x_i\})$ then the task is formulated in a different way: how will the distribution function $w(\{x_i\})$ be changed after binding part $\tilde{P}$ in a polycomplex with P?

Consider solution $\tilde{P}$ and distribution function $w_0(\{x_i\})$ normalized by the total base mole concentration of polymer m^0:

$$\int_{\{x_i\}} w_0(\{x_i\})\, d\{x_i\} = m^0 \tag{1}$$

Now add to this solution complementary polymer P which is capable of forming the equimolar (stoichiometry is not important since it must be only constant) polycomplex with $\tilde{P}$, its concentration being $m < m^0$, then part $\tilde{P}$ will bind in a polycomplex and the distribution function of $\tilde{P}$ $w_s(\{x_i\})$ remaining in the solution, normalized to $m^0 - m$, on the whole will not be identical to the initial one, i.e.,

$$\frac{1}{m^0 - m}\, w_s(\{x_i\}) \neq \frac{1}{m^0}\, w_0(\{x_i\}) . \tag{2}$$

Assuming that each macromolecule $\tilde{P}$ is either fully bound in a polycomplex or is fully in a non-bound state, the following correlation is valid for each fraction of polymer $\tilde{P}$:

$$\frac{w_{p.c.}(\{x_i\})}{w_s(\{x_i\})} = A \cdot \exp\left[-\Delta G(\{x_i\})/RT\right] \tag{3}$$

where $\Delta G(\{x_i\})$ is the free complex-forming energy of fraction $\{x_i\}$ of polymer $\tilde{P}$ with P. Hence, since

$$w_{p.c.}(\{x_i\}) + w_s(\{x_i\}) = w_0(\{x_i\})$$

then

$$w_s(\{x_i\}) = \frac{w_0(\{x_i\})}{1 + A \cdot e^{-\Delta G(\{x_i\})/RT}} \tag{4}$$

Parameter A denotes the normalizing multiplier, reflecting the fact that polymer P is in deficiency, i.e., that the polycomplex phase is of limited "holding capacity". If this capacity of a polycomplex is unlimited then, in accordance with Boltzmann's statistics, $A = 1$; therefore $A < 1$ in the given case.

Table 1. Investigated "Recognition" and Substitution Reactions

Type of reaction	No	Macromolecular partners			Comments	Ref.
		P	P_1	P_2		
1	2	3	4	5	6	7
"recognition" $P + P_1 + P_2 \rightarrow$	1	PMA	PVPD, PEO, PVA		Row of reactivity PVPD > PEO > PVA[1]	11)
	2	PMA	PEO, PAAm, PVPD, PVA		Row of reactivity PEO (MM 1.4×10⁶) > PAAm > PVPD > PEO (MM 3×10⁵) > PVA[1]	13)
$\rightarrow$ p.c. $(P \cdot P_1) + P_2$ or substitution	3	PMA	PVPD, PEO	PAAm	Row of reactivity PVPD > PEO > PAAM[1,2]	14)
	4	PAA	PVPD	PEO		15)
p.c. $(P \cdot P_2) + P_1 \rightarrow$	5	PMA	P4VP	P2Vp		13)
$\rightarrow$ p.c. $(P \cdot P_1) + P_2$	6	PEO	PMA	polyglutamic acid		13)
	7	p-brom-phenol-formaldehyde resin	PVPD	PEO		16)
	8	PVP-100	PVSA	PMA-Na		17)
Equilibrium	9	PMA	PVPD	PEO	← under MM PEO 15,000, MM PVPD 40,000 → under MM PEO 100,000, MM PVPD 5,000	18)
p.c. $(P \cdot P_1) + P_2 \rightleftarrows$	10	PVPD	PMA	copolymers AA-MA	← under growth of AA content in copolymers	19,20)
$\rightleftarrows$ p.c. $(P \cdot P_2) + P_1$	11	PVPD	PMA 6:41:53³	PMA 14:32:54³	← under growth of MM PVPD	
	12	PEO	PMA 14:32:54	PMA 6:41:53	← under growth of MM PEO	20,21)
	13	poly-2-diethyl-aminodextran	PVSA	polyglutamic acid	→ under growth of pH	22)
	14	PMA	PVPD	P2VP	← under pH = 3, → under pH = 6	13)
	15	PMA	PVPD	PEI	← under pH = 2, → under pH = 5	13)
	16	PVP-100	PPh-Na	PMA-Na	← under growth of MM PPh-Na ← under growth of concentration NaCl, LiCl → under growth of concentration KCl	

Notes: 1. In the row of reactivity, every next polymer is substituted with the preceding one in the polycomplex
2. Disagreement with the preceding rows are probably explained by the polymerization degrees of PVPD, PEO and PAAm which are not given in this work
3. Ratio of iso-, syndio- and heterotactic triads in PMA samples

Equation (7) shows that if P is in deficiency then the complex-formation process is accompanied with fractionation of $\tilde{P}$ by non-uniformity parameters. The efficiency of such fractionation or, in other words, the degree of "recognition" by polymer P of the optimum partner among $\tilde{P}$ will be determined by the type of function $\Delta G(\{x_i\})$. According to the available theoretical calculations, "recognition" (i.e., high fractionation efficiency) is a direct consequence of the reagents' polymer origin which determines high cooperativity of the system of bonds between macromolecules and strong dependence of the stability of the polycomplex on chain lengths; it is this consideration that has necessitated the inclusion of the following paragraph into this review.

2.1 Dependence of the Stability of the Polycomplexes on Chain Length

The dependence of the polycomplex stability on the degree of polymerization of the macromolecular reagents may be derived taking as an example the interaction between relatively short chains (conditionally called "oligomers") and relatively long chains of the second polymer reagent ("polymer"). The polymer chains may be viewed as one-dimensional adsorptive lattice if end effects are neglected, i.e. assuming that all macromolecules, present in the solution, are bound in one very long chain. Schematically the reversible adsorption of oligomers on a polymer appears as

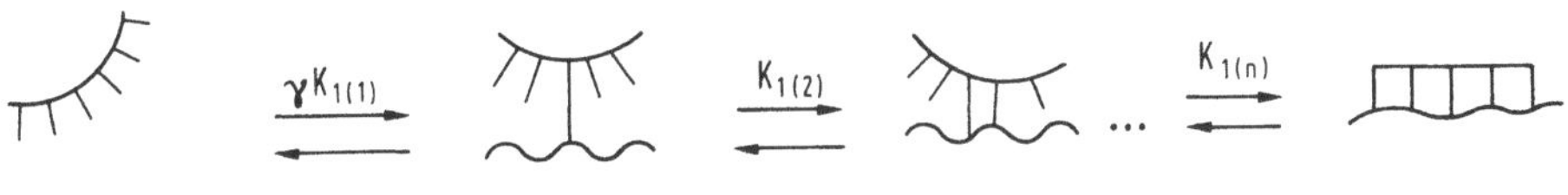

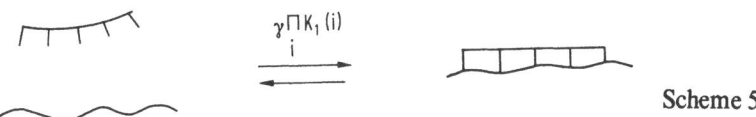

Scheme 4

where $\gamma K_{1(1)}$ is the probability of oligomer-polymer encounter and that of formation of the first bond between their units (γ depends on the oligomer and polymer concentrations, $K_{1(1)}$ — on the bond energy), $K_{1(2)}$, $K_{1(3)}$, ... $K_{1(n)}$ are the ratios of the probabilities of forming each next i-th bond and of disrupture of the (i — 1)-th bond, $K_{1(i)} = \exp(-\Delta g_i^0/kT)$ where n is the DP of the oligomer, Δg_i^0 is the free energy of forming the i-th bond under the existing (i — 1) bond.

According to the theoretical analysis of this model [24, 25], even with the single bond energy which only little exceeds that of the thermal motion, in practice an oligomer exists in equilibrium only in two states — free and completely bound, i.e., the interaction scheme being

Scheme 5

If polymer and oligomer are homopolymers, then the binding energies of all oligomer and polymer units may be assumed to be the same, i.e.,

$$\Delta g_i = \Delta g_1 \tag{5a}$$

$$K_{1(i)} = K_1 \tag{5b}$$

For instance, it has been shown for PEO (oligomer) — PMA (polymer) interaction that the reaction heat calculated per monomer unit does not depend on the DP of the oligomer [26]. Then the probability of oligomer-polymer association is

$$\gamma \cdot K_1^n = \gamma \cdot e^{-n \cdot \Delta g_1^0 / kT} \tag{6}$$

and the intermediate states of scheme (4) may be neglected, with consideration taken only of the initial and final states (Scheme (5)).

In this case, the equation of the isotherm of the adsorption of the oligomer on a one-dimensional lattice (polymer) under conditions

$$n \gg 1$$

$$\frac{\beta}{n(1 - \beta)} \ll 1$$

appears as

$$\frac{\beta}{m(1 - \beta)} e^{\frac{\beta}{1 - \beta}} = e^{-n \cdot \Delta G_1^0 / RT} \tag{7}$$

where β is the degree of "filling" the polymer with oligomer (β is equal to the fraction of monomer units of the polymer occupied by the oligomer), m is the free oligomer concentration in the solution (in base mole fraction), ΔG_1^0 is the free energy of interchain bond formation calculated per base mole of polymer or oligomer. Equation (7) cited in [27, 28], has been obtained using the method of calculating the thermodynamic probability of oligomers adsorbed on a one-dimensional lattice which was suggested in [29]. The right-hand side of Eq. (7) is the equilibrium constant of adsorption K_n of the oligomer whith DP = n:

$$K_n = K_1^n \tag{8a}$$

where

$$K_1 = e^{-\Delta G_1^0 / RT} \tag{8b}$$

A theoretical dependence of the equilibrium constant of Scheme 5 reactions on the length of macromolecules, analogous to (8a), was obtained by various methods as far back as in the 60s when deriving equations for calculating the energy of stacking interaction and estimating its role in complex formation of oligonucleotides with complementary polynucleotides [24, 25, 30, 31].

Experimental dependencies of the stability of polycomplexes on the length of macromolecules and temperature are, in general, satisfactorily described by Eqs. (7) and (8a) [28, 32]. However, there are a whole number of objective reasons owing to which these equations may be used for quantitative calculation of the thermodynamic parameters of interpolymer interactions as well as molecular "recognition" only if certain conditions are fulfilled. The thermodynamic parameters, thus obtained, represent some effective values.

Some of these reasons are:

(1) Stacking interaction between purine and pyrimidine end-groups of the oligo-nucleotides adsorbed on the polynucleotide chain (provided these end-groups are not separated by the free monomer units of the polymer). In such a case the adsorption equation can be analytically derived only for certain β, e.g., for $\beta = 0,5$ [30, 31] (an equation for any β has been obtained in one of the works [33] by Hill's method [34] although the binding of the monomer units of the oligomer in a chain was not taken into account):

$$\ln C = A_1 \cdot \frac{1}{T} + B_1 \qquad \text{at} \qquad n = \text{const} \qquad (9a)$$

$$\ln C = A_2 \cdot n + B_2 \qquad \text{at} \qquad T = \text{const} \qquad (9b)$$

$$\frac{1}{T} = A_3 \cdot \frac{1}{n} + B_3 \qquad \text{at} \qquad C = \text{const} \qquad (9c)$$

where C is the concentration of oligomer. These equations represent particular cases of Eq. (7) for $\beta = 0.5$.

(2) Changing flexibility of macromolecules under complexformation and changing energy of the interaction of their units with the solvent; as a rule, owing to screening of lyophilic groups in a polycomplex [1, 35], the solubility of polycomplexes is much lower than that of free macromolecules. It follows from Eqs (7) and (8) that white small energies of interaction A ... B are sufficient to form polycomplexes. Thus, apart from the bond energy $\Delta\varepsilon_1$, a considerable contribution to ΔG_1^0 can be made, for example, by the energy of long-range interaction $\Delta\varepsilon_2$ between units of a polycomplex in a poor solvent and by the energy of hydrophobic interactions $\Delta\varepsilon_3$, if the solvent is water, and so on, i.e. [1]

$$\Delta G_1^0 = \sum_i \Delta\varepsilon_i \qquad (10)$$

For example, in PEO or PVPD (oligomers) — PMA (polymer) systems the major contribution to stabilization of polycomplexes in water is made by hydrophobic interactions [26] (the formation reaction of these polycomplexes in water is endo-thermic). The relative contribution made by each of $\Delta\varepsilon_i$ to ΔG_1^0 varies with both changing reaction medium and changing degree of "filling" polymer with oligomer, i.e.,

$$\Delta G_1^0 = f(\beta) \qquad (10)$$

In some cases the value of ΔG_1^0 so rapidly increases with increasing of β that, with the average $\bar{\beta} < 1$, oligomer distribution takes place in accordance with the principle "all or nothing", i.e. for one part of the polymer $\beta \simeq 1$ and for the other one $\beta \simeq 0$ [36, 37]. This "self-organization" rather typical of interpolymer interactions with participation of short and long chains [38, 39].

(3) Defects in the structure of a polycomplex, such as different-type loops from the unbound-in-a-complex parts of macromolecules of type:

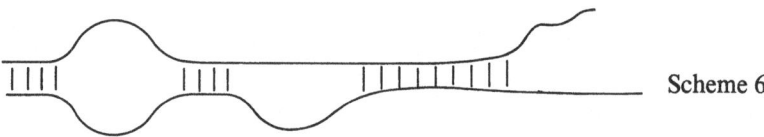

Scheme 6

particularly with considerable chain lengths of the components. Such defects may be present in the particles of a polycomplex, being, a kind of microreactor [40, 41], either due to equilibrium in the particle

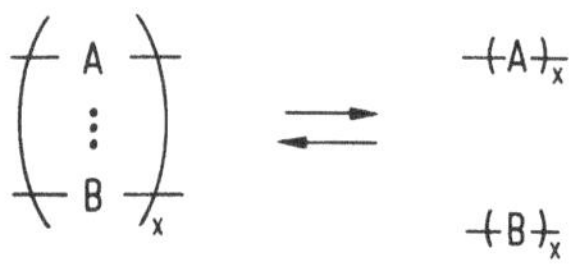

$$\left(\begin{array}{c} \text{+}A\text{+} \\ \vdots \\ \text{+}B\text{+}_x \end{array} \right)_x \quad \rightleftharpoons \quad \begin{array}{c} \text{+}A\text{+}_x \\ \\ \text{+}B\text{+}_x \end{array}$$

Scheme 7

or to incomplete reaction in the particles. The latter is caused by decreasing intra-molecular mobility of macromolecules during the development of continuous inter-molecular bond sequences after formation of primary intermolecular bonds which appear when the coils of complementary macromolecules come into contact with each other in solution. The presence of lengthy loops has been proved while investigating thermochemical reactions in polycomplexes [42–45]. The existence of reversible type (Scheme 7) reactions has been observed in polyelectrolyte complexes, particularly those with considerable chain lengths and relatively small fractions of inter-molecular bond θ in the particles of these complexes [41]; with small chain lengths of one of the components (oligomer), as shown in [30, 31], there develops a situation corresponding to Scheme (5).

The presence of defects caused by incomplete reaction in the particles has been found while comparing intra- and intermolecular reaction rates in polycomplexes obtained by mixing polymer solutions, and by matrix polymerization [46]. In the latter case a polycomplex is formed simultaneously with the chain growth which is connected with the complementary macromolecule, the matrix. This process of polycomplex formation is closer to the equilibrium one — in any case there are considerably fewer obstacles here for forming an uninterrupted sequence of inter-molecular bonds. That is why the rate and conversion of thermochemical reactions (which are connected with the presence or absence of defects — loops or tension in double-stranded chains of the polycomplex) depend on how the polycomplex have been obtained. After its destruction and reconstruction (e.g., by increasing and then decreasing of pH in the case of p.c. (PMA — PVPD)) the "matrix" polycomplex does not differ from the one obtained by mixing [46].

The presence of defects as such does not exert influence on type (7) or (8a) equations. A polycomplex with long macromolecules can be represented as a chain of Z segments sufficiently lengthy to consider the average total energy ΔG_s of intermolecular bonds as similar for all the segments. Then

$$K_n = e^{-Z \cdot \Delta G_s / RT}$$

and since Z is proportional to n, i.e., $Z \cdot \Delta G_s = n \cdot \Delta G^*$ then

$$K_n = e^{-n \cdot \Delta G^* RT} = K_1^n \qquad (11)$$

where $K_1 = e^{-\Delta G*/RT}$, ΔG^* and K_1 — corresponding effective values connected with the degree of nonperfection of the polycomplex structure which depends not only on the nature of interacting chains but on the conditions of the polycomplex formation and even on the length of the interacting chains as well.

Thus, the dependency type of the stability constant of the polycomplex on the chain length of the lowest molecular weight component, expressed by Eqs. (8a) and (8b), is correct. We would only like to note here that conclusions drawn by different authors might be erroneous when interpreting the experimental data on the dependence of the stability of the polycomplex on the length of macromolecules. For example, the rapidly growing dependence of the stability of the polycomplexes on the length of macromolecules under certain external conditions (pH of the medium, temperature, etc.) has stimulated the appearance in some works (e.g., [7]) of such definitions as "critical chain length" or "cooperative chain fragment", and so on. It is obvious that these definitions are misleading since changes in the stability of the polycomplex taking place along with the growing chain lengths are continuous; besides, the DP area of oligomer where stability is rapidly increasing with increase of n, depends on the conditions under which the reaction is running — temperature, concentration, medium; this follows from both Eq. (10) and the available experimental data.

Beside the "lower" one, a certain "upper critical length" of chains exists above which polycomplexes are found to be unstable [7]. This conclusion is based on the dependence of the viscosity of the solution of the mixture of two polymers on the molecular weight of one of them, with the molecular weight of the other

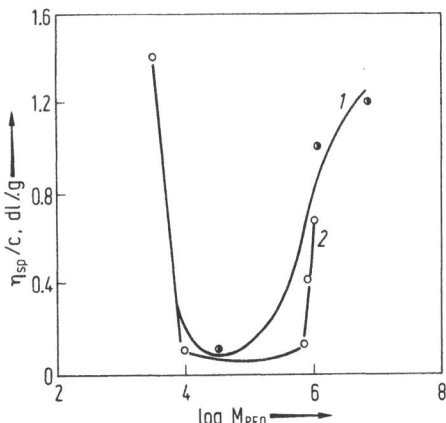

Fig. 1. Dependence of the specific viscosity of mixtures of the copolymer of the methacrylic acid and methylmethacrylate with PEO on the molecular weight of PEO. Total polymer concentration 0.05 g/dl, solvent — water:ethanol = 30:70; temperature 25 °C (*1*), 15 °C (*2*). [47, 7b]

polymer component being constant (see Fig. 1 from [47]). Decreasing solution viscosity of an equimolar mixture of PMA and PEO with increasing of length of the PEO chain in the range of low molecular weights is explained (like in earlier works [48, 49]) by the formation of a polycomplex and its growing stability while increasing viscosity in the range of high molecular weights is interpreted as decreasing polycomplex stability. The latter interpretation obviously rests on a misunderstanding because with the used concentration of PEO, the coils of low molecular PEO do not overlap and the forming particles of the polycomplex are individual and compact while the coils of high molecular PEO overlap (the solution becomes semidilute and the complex formation process is accompanied with structurization which, in fact, results in greater viscosity. It is noteworthy that the first report by Smith et al. [50] on complex formation in PAA and PEO solution was based on the data on the increasing solution viscosity of the polymer mixture at moderate concentrations.

As to Eq. (7), it is to be remembered that ΔG_1 in a general case is a function of β. Therefore, the experimental dependencies of β on concentration, chain length of oligomer and temperature may be employed to find thermodynamic parameters only for a fixed value of β, e.g., for $\beta = 0.5$ using Eqs. (8 a–b). These equations have been taken by various authors to calculate the enthalpy and entropy of complex formation between simple synthetic oligomers and polymers [28]. In a number of cases the correspondence between the values of complex formation enthalpy thus obtained and determined, either by calorimetry or by potentiometric titration [26], has been found satisfactory although it is obvious that in a general case these values do not necessarily coincide.

It will be shown below that type (11) equations directly lead to both the ability for molecular "recognition" and the limitations imposed on the mechanism of "recognition".

2.2 The Selectivity of Interpolymer Interactions with Regard to Chain Lengths. Fractionation by Molecular Weight

Consider the interaction of polymer P with polydisperse oligomer $\tilde{P}$ under the following conditions:

   (i) DP of P is greater than DP of the highest molecular weight fraction of $\tilde{P}$
   (ii) $\tilde{P}$ is in excess with regard to P
   (iii) under conditions of an experiment, the total degree of filling P with oligomers is equal to 1, i.e.,

$$\sum_i \beta_i = 1 \tag{12}$$

In this case, in Eq. (4) $\{x_i\} \equiv n$ and

$$\Delta G(n) = -nRT \ln K_1 \tag{13}$$

Then MWD of the non-bound (in a complex) oligomer boils down to the following equation

$$w_s(n) = \frac{w_0(n)}{1 + A \cdot K_1^n} \qquad (14)$$

Normalization parameter A is determined by the correlation of initial concentrations P and $\tilde{P}$, i.e., m and m^0, respectively.

In the general case of a polydisperse oligomer $\tilde{P}$, it is impossible to calculate the P-bound fraction analytically; according to [27], this task is solved on a computer if MWD of the oligomer, constant K_1 and the initial concentrations of oligomer $\tilde{P}$ and polymer P are given. An example of such a calculation is shown in Fig. 2a for the most probable distribution of the oligomer, K_1 which is close to K_1 for the poly-carboxylic acid-nonionogenic polymer systems, e.g., for the PMA-PVPD, PAA-PVPD, PMA-PEO, etc. systems [28].

For the sake of convenience, the distribution functions in Fig. 2a are normalized to the quantity of the oligomer; with such a normalization, the choice of the Y-axis scale is of no significance since it depends only on the total quantity of the oligomer; in this case the MWD changes of the oligomer when its fraction is bound in a poly-complex are more illustrative. It is clear From Fig. 2a that consecutive adding of the polymer to the oligomer solution results in consecutive narrowing of the free oligomer MWD at the expense of the high molecular weight part of the distribution bound in the polycomplex. If the polycomplex is removed, an oligomer fraction with rather narrow MWD may be extracted, the distribution width (dispersion) of this fraction depends on the portion of the polymer and on the value of K_1 [27]; the greater the value of K_1 and the smaller the portion of the polymer, the more efficient is the fractionation.

An analytical expression describing "recognition" in the mixture of polymer P and two monodisperse oligomers P_1 and P_2 with similar structures but different lengths n_1 and n_2, their initial concentrations being m_1^0 and m_2^0, respectively, was developed by Papisov and Litmanovich [27]. Provided conditions (i)–(iii) are fulfilled, the correlation degrees of filling chains P with oligomers P_1 and P_2 or, which is the same, the correlation of quantities of these oligomers, bound to polymer P, is expressed as

$$\frac{\beta_1}{\beta_2} = \frac{m_1^0 - \beta_1 \cdot m_p}{m_2^0 - \beta_2 \cdot m_p} K_1^{\Delta n} \qquad (15a)$$

where m_p is concentration of the polymer (all concentrations are expressed in base mole fractions, $\Delta n = n_1 - n_2$). The correctness of this equation does not depend on values β_1 and β_2 as, according to condition (iii), $\beta_1 + \beta_2 = 1$, i.e., the polymer chains are saturated with oligomers. In a particular case, if the oligomer concentrations are similar and just sufficient polymer is introduced to allow only half of the total amount of the oligomes to bind in a polycomplex (i.e., $m_1^0 = m_2^0 = m$) it follows:

$$\frac{\beta_1}{\beta_2} = K_1^{\Delta n/2} \qquad (15b)$$

It follows from equations (15a, b) that "recognition", i.e., selective binding of oligomers by a polymer, depends not on their degree of polymerization but only on the difference of these values. It is obvious that with $K_1 > 1$ and large Δn, the polymer "recognizes" only the longest chains in the mixture of such oligomers.

If macromolecules P_1 and P_2, competing for binding with P, are such that the chain lengths correlate as $n_1 > n_p > n_2$, where n_p is the degree of polymerization of polymer P (the term "oligomer" naturally cannot be applied to P_1), this is the case when the stability of p.c. $(P \cdot P_2)$ is determined by the length of chain P_2 and the

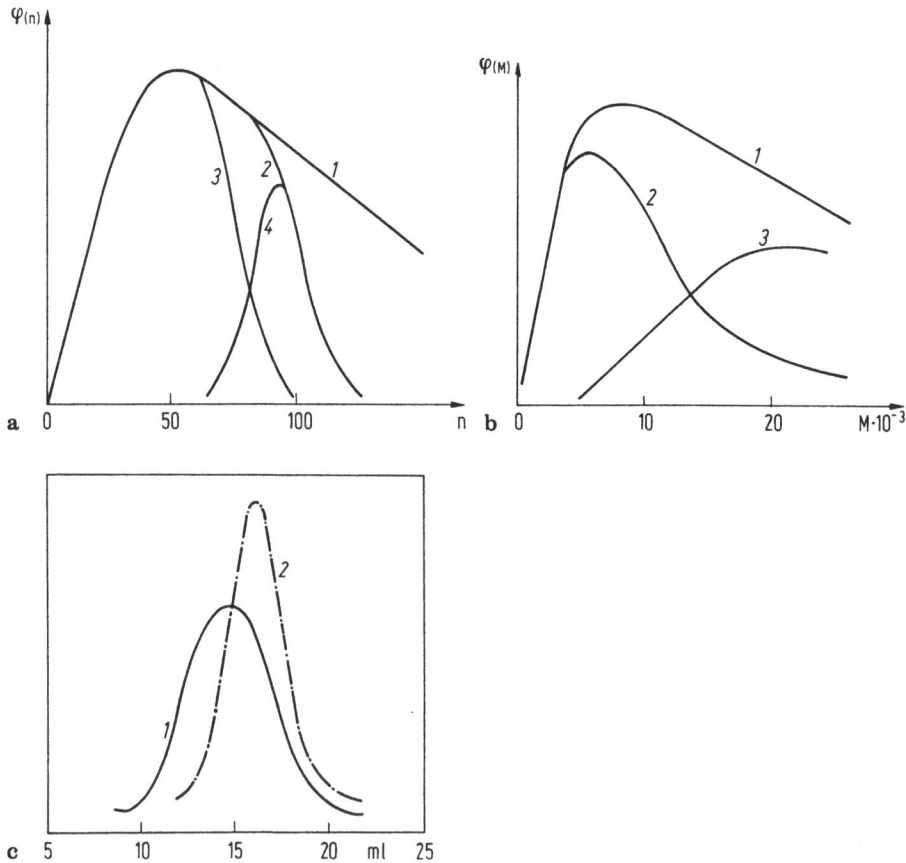

Fig. 2a–c. Changing MWD of polymers (oligomers) under complex formation. **a** Calculated after equation (14) for the initial $w(n) = 4.10^{-4} \cdot e^{-0.02 \cdot n}$ (1) with $\ln K_1 = 0.22$; (2) and (3) — MWD of the non-bound-in-a-polycomplex oligomer after addition the first ($m = 0.385m^0$) and the second ($m = 0.115m^0$) portions of the polymer, (1)   MWD of the oligomer bound in a polycomplex with the second portion of the polymer [27]. **b** (1) — MWD of the initial PVPD sample, (2) — MWD of PVPD remaining in the solution after binding in a polycomplex with PAA and removal of the polycomplex, (3) — MWD PVPD bound in a polycomplex with PAA [52]. **c** Fractionation of polymer PAA ($\bar{M}_w = 4.16 \cdot 10^5$) by oligomer PEO ($\bar{M}_w = 1.26 \cdot 10^5$); initial PAA sample (1) and that remaining in the solution after removal of the polycomplex (2), data obtained by GPC [53]

stability of p.c. ($P \cdot P_1$) by the length of chain P (because the stability is determined by the chain length of the component with the lowest molecular weight); then it may be shown that for case $m_1^0 = m_2^0 = m_p$

$$\frac{\beta_1}{\beta_2} = K_1^{\frac{n_p - n_2}{2}} \qquad (15\,c)$$

i.e., the efficiency of "recognition" depends on the difference between the lengths of chains P_2 and P but does not depend on P_1. The experimental data on separating a mixture of two polymers (P_1 and P_2) greatly differing in chain lengths through their complex formation with complementary polymer P with an intermediary chain length were obtained by Tsuchida and co-workers [13]. This work shows that only high-molecular weight PEO binds in a polycomplex after PMA with a molecular weight of $6 \cdot 10^5$ is introduced into the diluted solution of a mixture of the PEO samples with molecular weights $1:4 \cdot 10^6$ and $3 \cdot 10^3$.

The experimental data on fractionation by molecular weight are given in [52, 53]. Fig. 2b shows changes in the MWD of PVPD (oligomer) after binding part of it in a polycomplex with PAA (polymer) [52]; it is clear that the experimental pattern (Fig. 2b) corresponds to the theoretical one (Fig. 2a), i.e., the high-molecular fraction binds in a polycomplex while the low molecular one remains in the solution.

Similar results have been obtained by Kokufuta et al. [53] for the PEO (oligomer) — PAA (polymer) system although in this case terms "oligomer" and "polymer" are more conditional as their average molecular weights are close to each other and relatively high, i.e., $M_w = 1.26 \cdot 10^5$ and $M_w = 4.16 \cdot 10^5$, respectively. Binding part of polydisperse PEO (initial $\bar{M}_w/\bar{M}_n = 20.2$) in a polycomplex with PAA leads to considerable narrowing of the MWD of PEO which remains free in the solution. The efficiency of fractionation increases with increasing temperature. $\bar{M}_w/\bar{M}_n$ of PEO remaining in the solution goes down to 3.43 when the reaction takes place at $20°$ and to 2.46 at $50°$. This is in good agreement with the theory since the enthalpy of PEO—PAA complex formation in water is positive, the stability of the polycomplex, characterized by the value K_1, increasing with increasing temperature [26, 27]. The latter is connected with the considerable role of hydrophobic interactions in the stabilization of this polycomplex in water [26]. The efficiency of fractionation in water-alcohol media is expected to decrease with increasing temperature because the enthalpy of PEO-polycarboxylic acids complex formation becomes negative when water is replaced by water-organic mixtures [26, 28].

It was reported [53] that a PEO sample with $M_w = 1.37 \cdot 10^4$ and $\bar{M}_w/\bar{M}_n = 1.23$ is not fractionated by polyacrylic acid. Obviously, even if some change in MWD does take place upon the binding part of such a narrowly distributed polymer in a polycomplex, it is difficult to be estimated.

The available literature contains no theoretical analysis of the situation developing during fractionation of the polymer· by complex formation with the oligomer. According to Kokufuta et al [53], fractionation of PEO $\bar{M}_w = 1.26 \cdot 10^5$ by oligomer PAA $\bar{M}_w = 1.1 \cdot 10^3$ does occur although much less effectively that fractionation of an oligomer by a polymer. The fractionation of polymer PAA by oligomer PEO [53] is also possible (see Fig. 2c). The explanation of this possibility of fractionation is two-fold: either the stability constant of the polycomplex depends on the DP of both partners, which yet requires theoretic justification, or this fractionation is of the same nature as the fractional precipitation. Indeed, in the experiments of Kokufuta et al. [53] the polycomplex was separated from the excess of the free partner by adding HCl to the solution, upon which the polycomplex precipitated. It is known that in diluted solutions, particles of a polycomplex are practically dissociated [54], each one of them consisting of one polymer macromolecule and the corresponding number of oligomer molecules, i.e., the molecular weight of a particle of a polycomplex is determined by the molecular weight of the polymer. When a precipitating agent is added, the largest particles of the polycomplex primarily precipitate as usually upon fractional precipitation. It should be emphasized here that the "all or nothing" distribution principle does not work in cases with solutions of non-equimolar mixtures of polymer PAA and oligomer PEO [36], i.e., all macromolecules of the polymer are evenly filled with oligomers. Hence, the disproportion, i.e., free macromolecules of the polymer and its oligomer-saturated macomolecules appear only when the precipitating agent is added into the system; this is a strong argument in favour of the second mechanism of fractionation. Dispropotion in such systems is close to phase transition as was shown in [55].

2.3 Selectivity of Interpolymer Interactions with Regard to the Chemical Structure of Macromolecules

2.3.1 "Recognition" in a System with a Nonuniform Polymer; Fractionation by Composition

If, beside nonuniformity in chain length, polymer $\tilde{P}$ is also characterized by nonuniformity in composition, then to solve the molecular "recognition" task it is necessary to know the dependence of ΔG on the structure of the macromolecule.

To a first approximation, it is possible to take for the binary statistical copolymer with units A and B

$$\Delta G(a, b) = a \cdot \Delta G_{1a} + b \cdot \Delta G_{1b} \tag{16}$$

where a and b are the numbers of units A and B in a macromolecule of the copolymer, and ΔG_{1a} and ΔG_{1b} are the respective increments of the free energy. This additive approximation represents a natural generalization of Eq. (13) but it does not take into consideration a possible dependence of the complex formation free energy on the character of the distribution of monomer units in the polymer chain. Since no experimental data on such dependence are available at present, we will confine our analysis to the additive approximation which is valid in some systems. For instance, contributions of units of acrylic and methacrylic acids to the free complex formation energy of their copolymers with PVPD are additive (see 2.3.3). The equation describing the change of the function of the compositional and molecular weight nonuniformity of copolymers $\tilde{P}$ upon binding one of its part in a polycomplex with P, reads in the form of the additive approximation [20] as

$$w_s(a, b) = \frac{w_0(a, b)}{1 + A \cdot K_{1a}^a \cdot K_{1b}^b} \tag{17}$$

This equation is analogous to the equation describing fractionation of copolymers by fractional precipitation [56]. To obtain uniform fractions, it is necessary to use cross-fractionation [56, 57] in both cases since under a standard scheme of fractionation the macromolecules with different a and b but with similar denominator at the right-hand side of Eq. (17) will equally distribute between the solution and the polycomplex (between the diluted and the concentrated phases, respectively).

So far no direct attempt to test Eq. (17) experimentally has been described in the relevant literature. Clearly, such testing involves a number of methodological difficulties, in particular, the necessity for independent experimental determination of the functions of the compositional and molecular weight nonuniformity of the separated fractions and of the initial copolymer. In fact, cross-fractionation is perhaps the only possible direct method of experimental determining of the compositional and molecular weight nonuniformity of the copolymers; the algorithm for calculation of the nonuniformity by cross-fractionation was suggested in [58].

It has been shown [19, 20] that under PAA complex formation with excess of partly quaternized PVP, the latter is quite effectively fractionated by composition, macromolecules of the copolymer enriched with nonalkylated units being selected in the polycomplex. It is of importance that the preliminary molecular-weight fractioned sample of P-4-VP is quarternized with ethyl bromide, i.e., under the conditions of an experiment it is possible to consider

$$n = a + b \approx const$$

In this case PVP fractionation takes place only by composition and Eq. (17) becomes uniparametrical, analogous to equation (14):

$$w_s(\alpha) = \frac{w_0(\alpha)}{1 + A' \cdot \varkappa^\alpha} \tag{18}$$

where $\alpha = \dfrac{a}{a + b}$ is the fraction of quaternized units (composition of the copolymer) and $\varkappa = (K_{1a}/K_{1b})^n$. The dependence of the average composition of the separated fraction of copolymer on base mole correlation $\mu = [PAA]/[copolymer]$ is expressed [20] by equation

$$\bar{\alpha}_{p.c.} = \int_0^1 \frac{\alpha \cdot w_0(\alpha)}{\mu + A'' \cdot \varkappa^{-\alpha}} \, d\alpha \tag{19}$$

Table 2. PVP-49 Fractionation Under Interaction with PAA

μ = [PAA]:[PVP-49]	Average Compositions of PVP-q Fractions	
	In Polycomplex	In Solution
0.2	22	56
0.4	29	63
0.5	32	66
0.6	36	69
0.75	40	75

It is clear from Fig. 3 that this equation leads to a satisfactory description of the experimental data given in Table 2 using the distribution function by composition of the initial copolymer obrained from the data on the kinetics of quaternization [19].

According to Table 1, the greater part of experimental work was devoted to particular cases of selectivity of interpolymer interactions with regard to chemical structure (reactions of substitution). From the viewpoint of a theoretical description of such reactions, two cases should be distinguished: "polymer—two oligomers" systems and "oligomer—two polymers" systems.

2.3.2 "Recognition" in the Systems of "Polymer—Two Oligomers" Type; Inversion of Selection

Equations describing "recognition" by polymer P of optimum partner in the mixture of two oligomers P_1 and P_2 with degrees of polymerization n_1 and n_2 and effective constants of binding K_{11} and K_{12}, respectively, are obtained if to equation (17) is added the function of compositional and molecular wight nonuniformity

$$w(a, b) = \delta_{a, n_1} \cdot \delta_{b, 0} \cdot m_1 + \delta_{a, 0} \cdot \delta_{b, n_2} \cdot m_2 \tag{20}$$

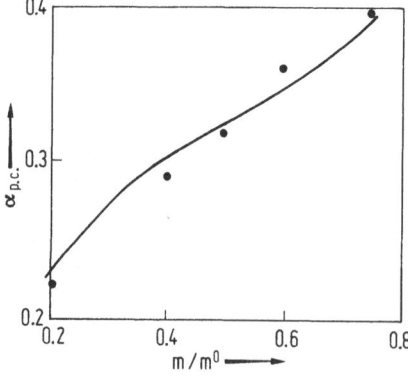

Fig. 3. Fractionation of PVP-49 under complex formation with PAA. *Dots* — experimantal data on the average compositions of fractions of the copolymer bound in a polycomplex with different quantities of PAA [19,20], *curve* — calculation from Eq. (19) with $\varkappa = 1.65$

where $\delta_{x,y}$ is Kronecker's symbol, $\delta_{x,y} = \begin{cases} 1 \text{ if } x = y \\ 0 \text{ if } x \neq y \end{cases}$, m_1 and m_2 are base mole

concentrations of P_1 and P_2. After elementary transformations, with account taken of conditions (i)–(iii) (see 2.3.1), it is possible to obtain

$$\frac{\beta_1}{\beta_2} = \frac{m_{1s}}{m_{2s}} \cdot \psi \tag{21}$$

where m_{1s} and m_{2s} are base mole equilibrium concentrations of free P_1 and P_2 in the solution of the polycomplex and ψ is the factor of selection ("recognition")

$$\psi = K_{11}^{n_1}/K_{12}^{n_2} \tag{22}$$

Equations (21) and (22) were obtained by statistical thermodynamic analysis [20, 27]. Fulfillment of condition (iii) means that Eq. (21) is valid for $(\beta_1 + \beta_2) \to 1$.

It is obvious that for $\psi \gg 1$, practically only P_1 binds in a polycomplex and for $\psi \ll 1$ only P_2, i.e., the equilibrium is practically completely displaced toward forming one of the possible polycomplexes and this was observed experimentally (see Table 1).

However, as follows from Eq. (22), the factor of selection depends on the DP of the oligomers, so under their different values, the polymer "recognizes" either one or the other oligomer or both. Abe, Koide and Tsuchida [13] first found experimentally that if low molecular weight PEO is a weaker complexing agent with regard to PMA than PVPD and PAAM then high-molecular weight PEO is a stronger one which substitutes PVPD and PAAM in their complexes with PMA. Analogous results for the PMA-PEO-PVPD system were obtained by Litmanovich [18].

Inversion of "recognition" (selection) also takes place under changing external conditions or even changing ratios of oligomer concentrations. K_{11} and K_{12} may depend on temperature, pH, ionic strength, etc., in a different way. Therefore, changes in these parameters bring about considerable changes in value of the factor of selection. The most trivial one is the case when changing external conditions result in the transition from the range where p.c. $(P \cdot P_1)$ is stable and p.c. $(P \cdot P_2)$ is not, to the range where p.c. $(P \cdot P_2)$ is stable but p.c. $(P \cdot P_1)$ is not.

Thus, polycomplexes stabilized by H-bonds are stable in an acid medium and decompose in a neutral one, and polycomplexes of weak polyacids (such as PAA or PMAA) with polybases are, on the contrary, stable in a neutral medium but unstable in an acid one. Therefore, along with pH change, an inversion of selection takes place (see examples 14, 15 in Table 1). This effect was used by the authors of work [59] for preparative separation of one of the macromolecular components from the polycomplex. A similar phenomenon was observed by Kikushi [22] for the polybase — weak polyacid — strong polyacid system (example 13 in Table 1). As the pH value increases the ionization degree of the weak polyacid increases, and hence its competitive power with the strong polyacid; accordingly, the portion of the weak polyacid in the precipitate of the polycomplex increases.

Kabanov, Zenin, Izumrudov and Bronich [23] found a much more interesting and non-trivial case of inversion of "recognition" (selection) in polyelectrolite complexes when changing the ionic composition of the solution in conditions where both the polycomplexes were stable (example 16 in Table 1). The state of equilibrium and the rate of its attainement (see Sect. 3) has been found to depend not only on the ionic strength of the solution but on the nature of the cation of low molecular weight salt as well. If in the presence of KCl, the equilibrium is practically totally displaced toward the polycomplex PMA-PVP, then it is displaced in the presence of NaCl and LiCl toward the polycomplex PPh-PVP, with total displacement of the equilibrium being achieved at different concentrations of these salts. It is almost one order of magnitude lower with LiCl than with NaCl (0.025 M and 0.2 M, respectively). The cause of the observed effects, as the authors [23] see it, is different affinity of polycomplex-bound polyanions to counterions of different nature. The equilibrium is displaced toward that polycomplex which decomposes at a higher concentration of the given salt [23]. However, the decomposing salt concentrations (0.3–0.5 M) were considerably higher than the ones under which displacement of the equilibrium was observed, i.e., the inversion of "recognition" (selection) occurred in the stability range of both polycomplexes.

Quantitative description of the system in question within a nonmodified model of unidimensional adsorption is impossible because, first, the forming polycomplexes are nonstoichiometric and, second, the polycation is an oligomer in relation to one polyanion (PMA) but a polymer to another one (PPh). It may be assumed, however, that the type of equation for the factor of selection in this system should not greatly differ from an equation of type (15 b, 22); in any case, it is to contain the factor $K_{12}^{n_2}$ (n_2 is the degree of polymerization of polyphosphate — an oligomer being in excess with regard to complementary polymer PVP), since in work [23] the inversion of selection ("recognition") was also observed on changing the degree of polymerization of polyphosphate.

2.3.3 "Recognition" of a Polymer by an Oligomer in the "Oligomer—Two Polymers" Systems

If oligomer P has a "choice" between macromolecules of two polymers — P_1 and P_2, then the degree of "recognition" by the oligomer chains is characterized by the ratio β_1/β_2 where β_i is the share fraction of units of polymer P_i which are filled with the oligomer. Naturally, the "recognition" in such systems may be spoken about if the oligomer in the system is in deficiency with regard to each of the polymers. This implies that no saturation of one of the polymer chains with the oligomer is possible in the presence of free chains of the latter; otherwise, the oligomer will bind to both polymers. Assuming that the effective constants of binding oligomer P with polymers P_1 and P_2 (K_{11} and K_{12}, respectively) do not depend on the degrees of their filling with the oligomer, i.e., $\Delta G_{1i} \neq f(\beta_i)$

$$\frac{B_1}{B_2} = \psi = (K_{11}/K_{12})^n \tag{23}$$

where n is the degree of polymerization of the oligomer and $B_i = \dfrac{\beta_i}{1 - \beta_i}$ $\exp\left(\dfrac{\beta_i}{1 - \beta_i}\right)$ [20,27]; Eq. (23) is obtained from Eq. (7) for each of the systems

$P - P_1$ and $P - P_2$ (the equilibrium concentration of the free oligomer in these equations is the same for both systems).

However, Eq. (23) cannot be used in a general case as $\Delta G_{1i} = f(\beta_i)$ and ψ usually depend on β_1 and β_2. Therefore, it was suggested in [20, 27)] that this equation is used for calculations in a limiting case when $(\beta_1 + \beta_2) \to 0$; then Eq. (23) is simplified:

$$\lim_{(\beta_1+\beta_2)\to 0} \beta_1/\beta_2 = \psi_0 = (K_{11}^0/K_{12}^0)^n \qquad (24)$$

where K_{1i}^0 are the corresponding constants relating to $\beta_i \to 0$. Thus,

$$\ln \psi_0 = -n \cdot \Delta\Delta G_0/RT \qquad (24a)$$

where $\Delta\Delta G_0 = \Delta G_{01} - \Delta G_{02}$ and ΔG_{0i} are the corresponding average free energies of interpolymer interactions calculated per mole of units for $\beta_i \to 0$.

It follows from Eq. (24a) that complex formation selectivity in "oligomer-two polymers" systems increases with increasing the DP of the oligomer and, at high n, leads to practically error-free "recognition" of the optimum partner even with insignificant difference in the free energies of binding $\Delta\Delta G_0$. Such "recognition" has been experimentally observed in reality (examples 10–12 in Table 1).

Quantitative testing of Eq. (24, 24a) was carried out in works [19–21)] taking an example of selective interaction with participation of structurally not very differing macromolecules — copolymers and stereoisomers. The results are presented in Figs. 4, 5.

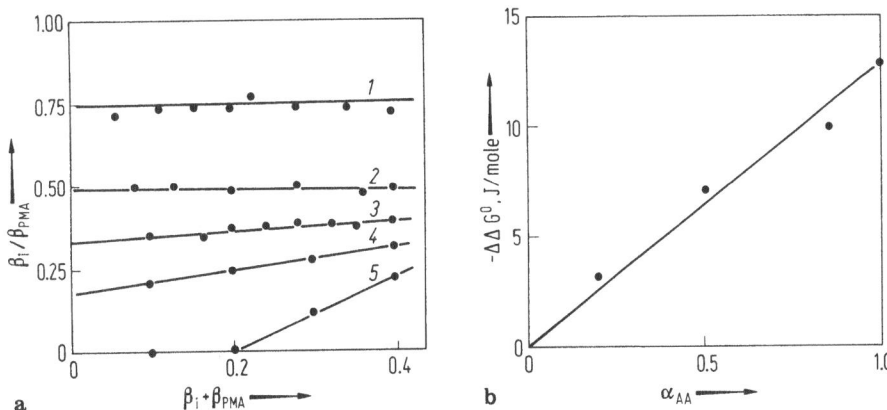

Fig. 4a. Dependence of the degrees of filling PMA (β_{PMA}) and copolymers AA-MAA (β_i) with PVPD oligomer on the total degree of filling of the polymer with oligomer. Content of units AA in copolymers: 20 (*1*), 50 (*2*), 85 (*3, 5*), 100 (*4*) mole %; n_{PVPD}: 80 (*1–4*), 360 (*5*). [19, 20)] **b.** Dependence of the difference between average complex formation free energies per base mole, calculated by Eq. (24a), on the composition of copolymer $\alpha_{AA} = \dfrac{[AA]}{[AA] + [MA]}$

PMA is selected by oligomer PVPD from a mixture of PMA and copolymer AA-MA, the efficiency of "recognition" increasing along with increasing content of AA units in the copolymer and growing degree of polymerization of PVPD (Fig. 4a). $\Delta\Delta G_0$ value calculated with Eq. (24a), depends linearly on the copolymer composition (Fig. 4b) which means that contributions to the free complex formation energy of units AA and MA are additive.

Exceptionally high selectivity of intermacromolecular reactions is illustrated by the fact that even an insignificant difference in the microtacticity of two PMA samples (i:h:s = 6:53:41 and 14:54:32, respectively) is found to be sufficient for oligomers PVPD and PEO to be able to "recognize" one of them (Fig. 5a). It is noteworthy that oligomers prefer different stereoisomers: PVPD preferably binds the sample enriched in *iso*tactic triads, and PEO that bind *syn*diotactic triads. If the "recognition" were conditioned by differences in the thermodynamic potentials of the free-state stereoisomers, then the same sample would be selected by both PEO and PVPD. Therefore, it seems reasonable to consider the structural correspondence of the components of the polycomplex to be the main cause of selection in the given case.

As shown in Fig. 5a, the efficiency of selection increases with the growing degree of polymerization of PEO, the corresponding dependence in the coordinates of Eq. (24a) being linear (Fig. 5b). Calculated from the tangent of the straight line, the difference between the average complex formation free energies of PEO and the given stereoisomers PMA for $(\beta_1 + \beta_2) \to 0$ comprises only 42 J/base mole. Values of the same order determine "recognition" in the case of copolymers AA-MA as well (Fig. 4b).

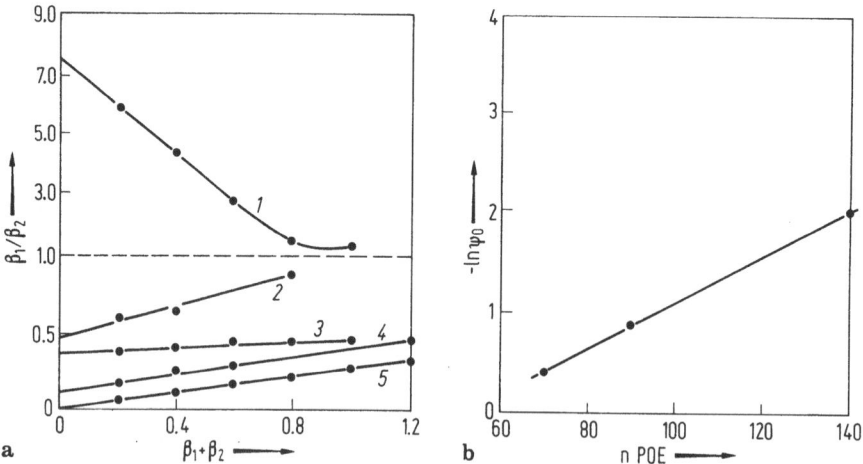

Fig. 5a. Dependence of degrees of filling two PMA samples of different microtacticity with oligomers PVPD (*1*) and PEO (*2–5*) on the total degree of filling of the polymers. β_2 — the degree of filling the PMA sample with I:H:S = 14:54:32, β_1 — the same for a sample with I:H:S = 6:53:41. The oligomer polymerization degrees: 80 (*1*), 70 (*2*), 90 (*3*), 140 (*4*), 450 (*5*)[21]. **b** Dependence of the characteristic value of selection factor ψ_0 on n for selective interaction of PEO with PMA stereoisomers in the coordinates of equation (24a)

Clearly, such a negligible difference in the interaction energies is capable of manifesting itself only in interpolymer reactions but never in reactions between low molecular compounds.

Found in a number of works [27, 60, 61], the distribution of oligomers among the chains of polymers according to the "all or nothing" principle under deficiency of the oligomer in the reaction system may be viewed in a different light as the basis of the above argumentation. Since every polymer is to be characterized by greater or lesser nonuniformity in chain microtacticity, the above distribution may be determined by the fact that the oligomer selects fractions of the polymer with a certain structure. Thus, along with other possible causes of self-organization in "oligomer-polymer" systems as manifested in the distribution of oligomer chains according to the "all or nothing" principle [28, 38, 39, 61], a significant role may be played by the compositional nonuniformity of the polymer and, hence, the molecular "recognition" of macromolecules with a certain consequence of isomer units.

3 Mechanism of Molecular "Recognition" and Kinetics of Interpolymer Substitution and Exchange Reactions

As has been shown in the previous sections, high selectivity of interpolymer interactions represents a fundamental property of macromolecular systems. This selectivity follows from the exponential dependence of the stability of polycomplexes on chain length (Eq. 11) and works independently of the nature of macromolecular bonds. A question then arises: what mechanism makes these reactions run if, at considerable chain lengths, value of the stability constants of the polycomplexes become so great as to result in the practical absence of the dissociation of these polycomplexes into free components. For example, for system PMA — PEO $\Delta G_1 \simeq -250$ J/base mole [26], hence, if their chain lengths are n > 1000, then $K_n > 10^{40}$ but that has no real physical sense and means only that the probability of the dissociation of the polycomplex equals zero.

On the other hand, molecular "recognition" in mixed polymer solutions may take place, as pointed out above, only through multiple "trials and errors" (see Scheme (3), i.e., the necessary condition for "recognition" is substitution of weaker macromolecular complexing agents in polycomplexes by stronger ones. The fact that interpolymer reactions of "recognition" and substitution are capable of running at great rates even at high molecular weights of the polymer means that there must exist a mechanism of substitution not requiring preliminary dissociation of the polycomplex formed with participation of the macromolecule being substituted.

This stimulated Kabanov, Papisov et al. [11] to put forward an assumption that depending on the chain lengths, interpolymer reactions of substitution may be run via two mechanisms which are at present commonly called "dissociative" and "contact" mechanisms.

At small lengths of chains P or P_1 and P_2, the dissociative mechanism is possible as described by Scheme 8

$$\text{p.c. } (P \cdot P_1) + P_2 \rightleftarrows P + P_1 + P_2 \rightleftarrows \text{p.c. } (P \cdot P_2) + P_1 \qquad \text{Scheme 8}$$

i.e., chains P_2 interact with free chains P present in the solution due to the reversibility of the formation-dissociation of p.c. $(P \cdot P_i)$ reactions.

At large lengths of chains, p.c. $(P \cdot P_1)$ practically does not dissociate, and then the contact mechanism comes into play through the interaction of P_2 with the free sequences — loops of macromolecules P in p.c. $(P \cdot P_1)$ and development of "zip-up" type substitution within a three-chain particle:

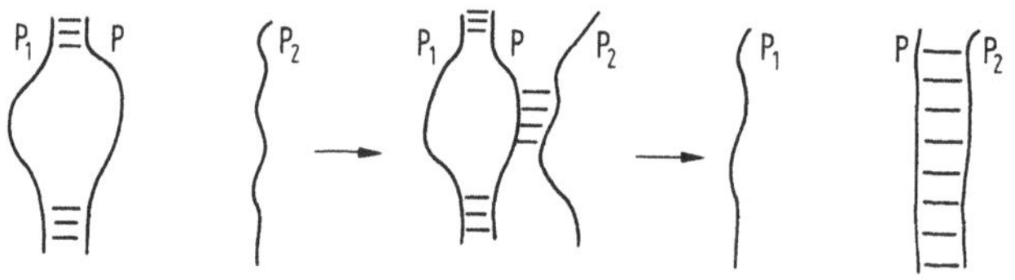

Scheme 9

i.e., p.c. $(P \cdot P_1) + P_2 \rightarrow$ p.c. $(P \cdot P_1 \cdot P_2) \rightarrow$ p.c. $(P \cdot P_2) + P_1$

Confirmation or rejection of these mechanisms were looked for in studies concerned with kinetics of interpolymer reactions of substitution and exchange. Results of the first investigation were published in 1977 [62]; in recent years, there has appeared a whole series of works in kinetics of reactions of substitution and exchange in hydrogen-bond stabilized polycomplexes [63, 64] and in polyelectrolyte complexes [17, 65–70]. The kinetics of the reactions was observed by optical methods: polarized luminescence [62], quenching of luminescence [17, 65–70] and fluorescence [63, 64]. The idea of these methods is the introduction into the composition of one of the reaction participants (polycarboxylic acid) of a small quantity of a marker whose spectroscopic parameters depend on whether this component is part of the polycomplex or is in a free state. For instance, the value of luminescence depolarization of an anthryl marker depends on the intramolecular mobility of the chain part to which it is connected, and sharply decreases when the polycomplex is formed [62]. The same marker was used by the authors of works [17, 65–70]; here the units of the second macromolecular partner, PVP, were quenchers of luminescence, therefore, the complex formation-dissociation processes with participation of marked PMA and PVP were accompanied by quenching-inflammating of luminescence. In works [63, 64], a danzyl marker was used the fluorescence intensity of which depends on its micromedium, and the fluorescence intensity of which in a polycomplex (hydrophobic medium) is higher than in the free state. The main results of the works under discussion are summarized in Table 3.

Both mechanisms of substitution were confirmed by the data of the very first investigation on the kinetics of interpolymer reactions of exchange [62]

$$\text{p.c. } (P \cdot P_1) + P_1^* \rightleftarrows \text{ p.c. } (P \cdot P_1^*) + P_1 \,, \qquad\qquad \text{Scheme 10}$$

where P is PEO or PVPD, and P_1 and P_1^* are PMA or PAA macromolecules, without and with luminescent markers, respectively. The rates of the exchange reactions in these systems were found to be much lower than those of the substitution reactions which gives the possibility of measuring the rates of exchange reactions by the polarized luminescence method.

Table 3. Kinetics of Exchange and Substitution Reactions

Type of Reaction	No	Macromolecular Partners			Factors Influencing Rate of Reaction			Ref.
		P	P_1	P_2	accelerating	slowing down	no influence	
1	2	3	4	5	6	7	8	9
Exchange								
p.c. $(P.P_1)$ + P_1^* →	1	PEO, PVPD	PAA, PMA		growth pH, PPA > PMA	growth MM PEO before 6,000	growth MM PEO after 20,000	62)
	2	PEO	PAA		growth pH, T, concn of polymers	growth concn NaCl	growth MM PEO from (24–77) × 10³	63)
→ p.c. $(P.P_1^*)$ + P_1	3	PEO, PVPD	PAA					64)
	4	PVP-100	PMA		growth concn NaCl		MM PVP-100 from (200–265) × 10³	65–67)
	5	copolymer γ,γ-dipyridyl and dibrom-propane	PMA			growth MM of copolymer		68)
	6	PVP-q	PMA		growth concn NaCl	growth q	MM PVP-q under fixed q	69)
	7	PVP-100						70)
Substitution	8	PVP-100	PVSA	PMA	growth concn NaCl		MM PVP-100	17)

It was shown that the rate of the exchange reaction increases as the poly-complex stability decreases (with increasing pH, under substitution of PMA by PAA) as well as the chain length of the lowest molecular weight component (oligomer). Some of these effects were later reproduced by Chen and Morawetz [63] — see Table 3. Fig. 6 shows the influence exerted by the chain length of the oligomer on the rate of the exchange reaction; $\varkappa = \bar{\beta}_t^*/\bar{\beta}_\infty^*$ is the degree of the reaction's complete-ness during a certain period of time t, $\bar{\beta}_t^*$ and $\bar{\beta}_\infty^*$ are the average degrees of filling the marked polymer after time t on mixing and in the equilibrium state (with $t \rightarrow \infty$), respectively. The rapidly decreasing rate of exchange with increasing chain length of the oligomer in the range of relatively small n is related by the authors [62] to the increasing stability of the polycomplex. The very fact that the exchange reaction runs as well as the practical independence of its rate on the chain lengths in the range of large n is in agreement with the contact mechanism Scheme 9. Thus, the transition from one mechanism to the other takes place along with changing stability of the polycomplex, i.e., its ability of dissociation; this transition may be connected with any factor exerting influence on the poly-complex stability such as changes in the chain lengths of the components, the reaction media, and the temperature.

Since the above discussed mechanisms of interpolymer substitution reactions rest on the most general peculiarities of the structure and properties of the polycomplexes, they must be common for reactions with participation of polycomplexes stabilized by interactions of any nature such as hydrogen bonds, electrostatic, etc. This is confirmed by numerous studies the results of which have been published in recent years.

Indeed, the rate of the exchange reactions in the polyelectrolyte complex PMA — copolymer (γ, γ — dipyridyl and dibromopropane) considerably decreases with increasing of DP of the copolymer in the range of small n [68]. In the range of high degrees of polymerization, the rates of the exchange reactions in poly-electrolyte complexes of PMA with partly or completely quaternized PVP do not

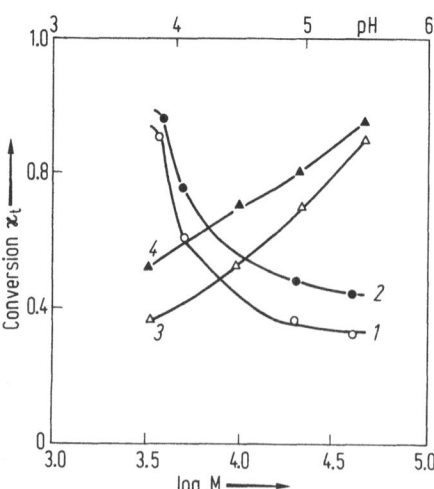

Fig. 6. Dependences of conversion degree of the exchange reaction in the system p.c. (PMA-PEO)-PMA* on molecular weight of PEO (1, 2) and on pH (3, 4) during time 10 (3), 100 (1) and 10000 (2, 4) minutes [62]

depend on n of the polycations [65-67,69]; nor does the rate of the substitution reaction of PMA with polyvinylsulfonic acid in the polycomplex PVP-100-PMA depend on n of PVP-100 [17].

The rate of interpolyelectrolyte exchange reactions also quite significantly depends on the polycomplex stability. For instance, it decreases, all other conditions being equal, with the growing charge density on the macromolecule of the electro-lyte as demonstrated in [69] for PMA and PVP complexes with different degrees of alkylation (under experimental conditions where the nonalkylated units of PVP are not charged). The stability level of most polyelectrolyte complexes is so high that while not running at all or running at a very low rate in the absence of low molecular weight salts [17,65-67,69], the substitution and exchange reactions are accelerated and are completed literally within seconds — with increasing ionic strength [17,65-67,69]. In contrast to polyelectrolyte systems, the rate of the ex-change reactions in hydrogen-bond stabilized polycomplexes decreases as the ionic strength increases [63].

A direct proof of the existence of the contact mechanism Scheme 9 was obtained by the authors of work [71] who used ultracentrifugation to observe a fraction consisting of all three polymers, i.e., a triple polycomplex which was an inter-mediary (and in some cases the final [72]) product of the substitution reaction.

Processing of the available experimental data with respect to kinetics of ex-change reactions (see Fig. 7a, b) shows that two steps may be singled out in this reaction — fast and slow ones, — each of them exhibiting a linear dependence in the first order of kinetics coordinates. Even though the use of the first-order coordi-nates is groundless (the exchange has been recently proved to follow the second order [73]), the rapid step can be correlated to the formation of the intermediary

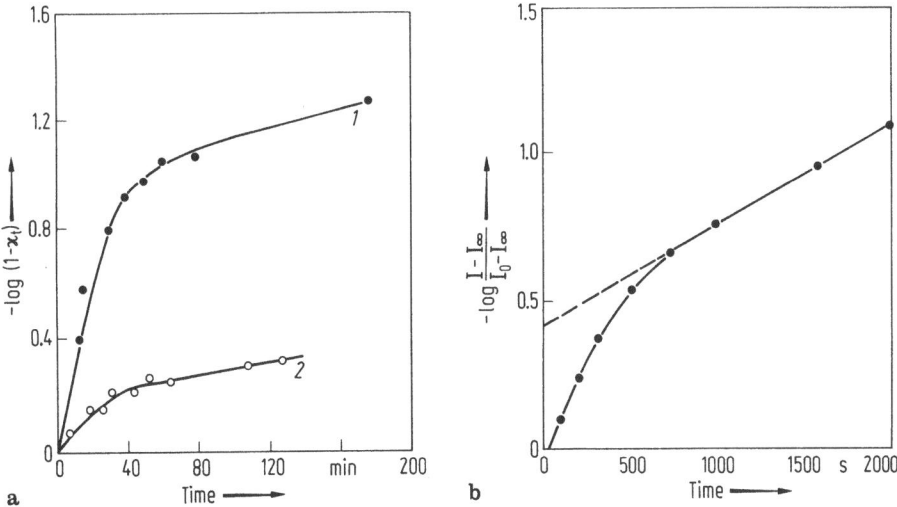

Fig. 7a. Kinetics of exchange in system p.c. (PMA-PEO)-PMA* in the first-order kinetics coordinates under the initial base mole ratio PEO/PMA = 0.5 (*1*) and 0.9 (*2*) [62]. **b** Kinetics of exchange in system p.c. (PAA-PEO)-PAA* in the first-order kinetics coordinates [63]

particles of the triple polycomplex, and the slow step to their further evolution. The rate constant of the exchange reaction is independent of temperature (found by the authors of this work) and shows that the basic contribution to ΔG^* is made by the entropy factor. The latter conclusion cannot be spread to all poly-complex systems since, for example, the rate constant of the exchange reaction in polycomplex PAA-PEO does depend on temperature [63].

Oparin, Gladilin et al. [8-10, 74] demonstrated that it is possible for substitution reactions to run in dispersions of insoluble polycomplexes for a broad range of systems. The interference-phase contrast method was used to observe under a micro-scope two types of morphological changes in the initial coacervate drops of insoluble complexes as the substitution reaction progressed. When the polynuclectide (polyadenylic or polycytidylic acid) in its complex with protamine is substituted with polyphosphate, there takes place "dissociation" of the initial coacervate drops, i.e., their fragmentation into smaller ones. When histon in its complex with gum arabic is substituted with polylysine, "vacuoles", i.e., zones with lower concentration of polyelectrolytes form in the initial coacervate drops. In the author's opinion, the obtained results seem to be of significance for the understanding of possible directions of the prebiological evolution and some intracell processes. The data in the works under consideration point to a very interesting and non-trivial peculiarity of interpolymer substitution reactions: the ability of the substituting macromolecule not to remain at the particle's surface, i.e., the coacervate drop (which is exactly what would be expected to happen), but to penetrate rather rapidly into the particle; the very formation of the new polycomplex particles within the primary ones points exactly to that. Obviously, the mechanism of involving parts of the new polycomplex into the primary particles is of considerable interest. Probably, it is a lesser compatibility of these parts with the solvent that enables such substitution reactions to run within a short period of time.

4 The Role of Molecular "Recognition" in Matrix Polymerization and Polycondensation

In a general case (no exotic cases of polymerization by the "zip-up" mechanism of preliminarily matrix-bound molecules of the monomer are considered here), matrix polymerization runs through consecutive addition of monomer molecules, reversibly adsorbed at the macromolecular matrix, to the active end of the growing ("daughter") chain which is also matrix-bound by intermolecular interaction [1, 4]. Under matrix polycondensation, the growing of the "daughter" chains proceeds by binding of monomers, growing oligomers and polymer chains adsorbed on the matrix.

In both cases, cooperative interaction between the growing chains and matrices is necessary for the completion of the matrix process. The necessity of the matrix — growing chain polycomplex is the cause of several molecular "recognition" connected peculiarities of matrix polymerization and polycondensation. This imposes limitations primarily on the chain lengths of the matrix and the "daughter" macromolecules.

4.1 "Recognition" of the Matrix by the "Daughter" Chain

Since no information can be passed from the matrix to the "daughter" chain without holding the growing chain on the matrix, the matrix polymerization takes place only provided the chain lengths of the matrices and the "daughter" chains are not shorter than is necessary to form a stable polycomplex between them.

According to Osada, Papisov, Kabanov et al. [75, 76], under polymerization of AA and MA in the presence of PEO, the kinetic matrix effect of PEO (decreasing rate of polymerization) is observed only if the molecular weight of the matrix-PEO — is greater than a certain value (Figs. 8a, b). It is clear from these pictures that the chain lengths of the matrix in every case practically coincide with those of PEO under which, as its molecular weight increases, PEO begins to interact with the preliminary prepared polymers — PMA and PAA. The solution's viscosity and the pH value serve as the indicator of polycomplex formation since particles of polycomplexes PEO-PAA and PEO-PMA in water are compact as compared to free molecules [76].

It follows from Eq. (12) that the growing chains are capable of forming a sufficiently stable complex with the matrix, i.e., of "recognizing" it, only after having reached

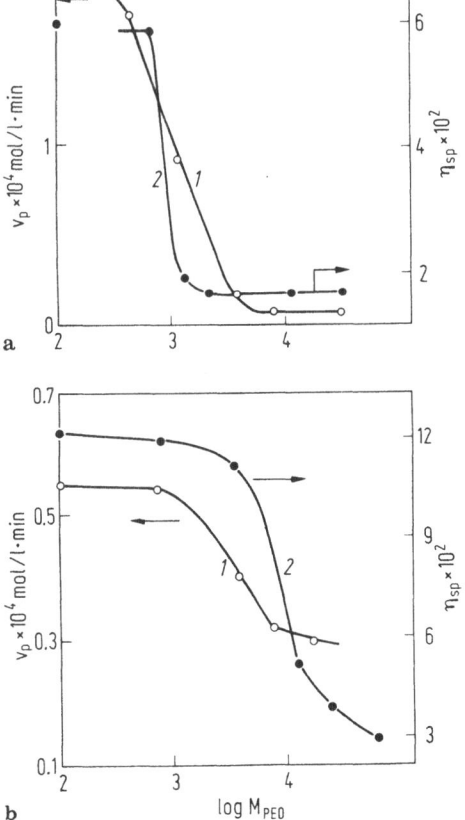

Fig. 8a. Dependence of the polymerization rate of MA on the molecular weight of the matrix — PEO (*1*) and of the specific viscosity of equimolar mixture PMA-PEO on the molecular weight of PEO (*2*) [76]. **b** The same for systems AA-PEO (*1*) and PAA-PEO (*2*) [76]

a certain length (more exactly, interval of lengths) in the process of matrix-independent growth [75, 76]. It is only after this "recognition" that the matrix starts controlling the elementary acts of the growth of the "daughter" chain. In other words, "recognition" of the matrix by the growing chains is a necessary condition for matrix synthesis [1].

The degree of polymerization of the growing chains at which they begin to "recognize" the matrix can be measured, if the matrix exerts influence on the rate of monomer conversion (kinetic matrix effect); it is necessary to find the concentration of the transfer agent at which the kinetic matrix effect disappears, and to determine the molecular weight of the polymer formed. For example, it was found with this method that for MA polymerization, the growing chains "recognize" the matrix — PEO macromolecules — when the PMA chain length reached about 10–15 monomer units [77].

It is noteworthy that in the above-considered PMA-PEO system, under comparable conditions, the chains of oligomer PEO "recognize" the chains of polymer PMA at a degree of polymerization of the oligomer of about 40–50 monomer units. The reason for that might be that high molecular PMA in water solution is present in a specific conformation stabilized by intramolecular hydrogen bonds and hydrophobic interaction [78]. Polycomplex formation between PMA and PEO is accompanied with destroying this specific PMA conformation [79] which requires additional energy; complex formation of PEO with short chains of PMA takes place without additional energy consumption. Hence, the average free energy of polycomplex formation, calculated per mole of units ΔG_1 in Eq. (7, 8b), and, accordingly, the value K_1 in Eq. (8a, 12) is found to be dependent on which of the polycomplex components is the oligomer. K_1 turns out to be greater in the PMA-PEO system if PMA is the oligomer.

An analogous asymmetry was recently found by Iliopoulos and Audebert [80] for the polymer PAA-oligomer PEO and polymer PEO-oligomer PAA systems. They offered a more general explanation not requiring specific conformation in one of the polymers. According to their analysis of the above-mentioned systems, the entropy losses under polycomplex formation are greater in case of more flexible oligomers (PEO). Hence, constants K_1 will necessarily depend on which of the polycomplex components is the oligomer; this appears to be a sufficiently general phenomenon.

The very fact that, after the "recognition", the growing chain is firmly held by the matrix, implies the possibility of any rate of the matrix process — it can be higher, lower or equal to that of polymerization with the matrix absent [1].

If all the "daughter" chains, as they grow, "recognize" the matrix macromolecules and grow further to higher length under their control, the rate of matrix polymerization may be independent of matrix concentration, i.e., zero-order kinetics with respect to matrix concentration is observed. This was found, for example, for the matrix zwitter-ion polymerization of 4-vinylpyridin on polyacids [81], and for the matrix radical polymerization of AA and MA on PEO [76]. At the same time, the dependence of the rate of the radical polymerization MA on P2VP matrix on concentration of P2VP is of a much more complex nature (see Fig. 9 in work [82]) and to explain it requires studying the preferred absorption of the monomer molecules at the matrix. With increasing matrix/monomer ratio at fixed concentration

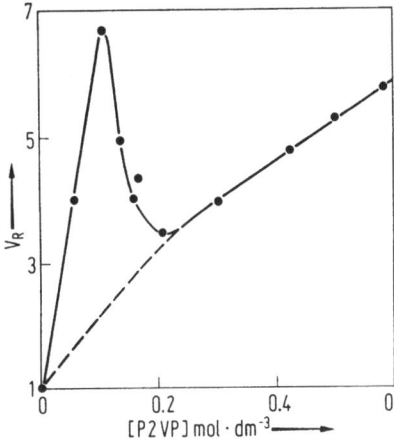

Fig. 9. Dependence of the relative polymerization rate of MA on the concentration of the matrix, P2VP [82]

of the monomer, the adsorption of the latter at the matrix decreases but the fraction of matrix-caught growing chains increases. Joint action of these factors determines the presence of the maximum on the curve in Fig. 9, since the rate of chain-growth on the matrix depends on the concentration of both the adsorbed molecules of the monomer and the adsorbed growing chains [82].

The kinetic scheme of the matrix polymerization, which takes into account adsorbtion of both monomers and growing chains at the matrix, was suggested by Smith, Tan and Challa in [83]. They used this scheme to successfully describe the experimental data on the dependence of the rate of polymerization of methacrylic acid at a P2VP matrix on the concentration of the initiator; they obtained a theoretical value for the order of the reaction of 1/4 in initiator which agrees well with the experimentally found value of 0.26.

4.2 Influence Exerted by "Recognition" on the Structure of "Daughter" Chains, Structure and Properties of Polycomplexes — Products of the Matrix Reactions of Polymer Synthesis

Under matrix synthesis in living organisms, the structure of matrices such as nucleic acids unambiguously determine the sequence of monomer units in the "daughter" macromolecules, DNA, RNA, proteins. Under matrix polymerization and polycondensation in much more simple systems, matrices are also capable — to a greater or lesser degree — of exerting influence on the composition and structure of the "daughter" chains.

In a very simple case, matrices play the role of microreactors creating a reaction medium near the active ends of the growing macromolecules which differs from that in the rest of the solution. For example, for matrix polymerization of MA on PEO macromolecules, the stereostructure of the PMA "daughter" chains is similar both in water and in benzene: in fact, it is the same as for polymerization of MA without matrix in alcohol media [24]. Probably, this is determined by the fact that the dielectric permetivity in the microreactor, where the active centre of the growing chain is situated, differs from that of water and benzene, being close to the dielectric permitivity of the alcohol medium.

Under copolymerization, the preferred sorption of one of the monomers on the matrix leads to enriching the "daughter" chains with this monomer. For instance, Polovinsky showed that MA — methyl methacrylate [85] and MA-styrene [86] copolymers, formed in the presence of a PEO matrix, are enriched in MA, the constants of copolymerization depending on the concentration of the matrix. It is shown by Kargina et al. [87] that in the presence of a PPh-Na matrix copolymers of 2-methyl-5-vinylpyridine and AA are considerably enriched in the cation monomer up to complete elimination of AA and formation of homopoly-2-methyl-5-vinylpyridine.

Polovinsky [86] gives equations describing the copolymer composition for matrix polymerization obtained with reversible binding of the active ends of the growing chain with the matrix taken into account:

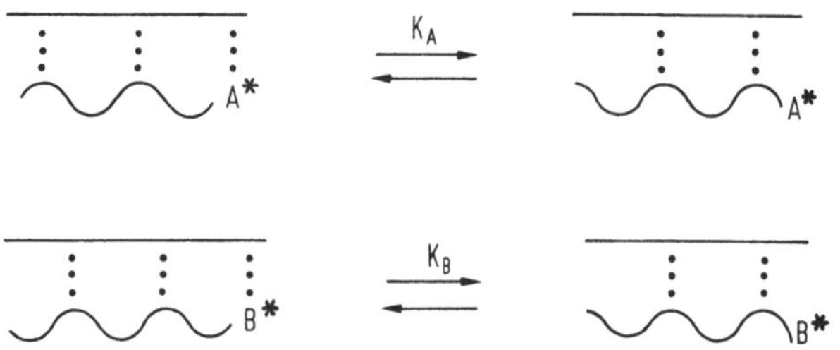

Scheme 11

The influence exerted by the matrix on the direction of the elementary growth steps of the "daughter" chains was observed for the matrix polymerization of 4PV on polyacids when the "daughter" P4VP had ionene structure [81], and for the matrix polycondensation of urea and formaldehyde in water, with PAA being present [88, 89]. In the latter case, the "daughter" chains of PFU contained the structures

$$-CH_2-N- \\ \quad\quad | \\ O=C-NH_2$$

Scheme 12a

which are practically absent in standard PFU since the activity of the mono-substituted amide group in the condensation reaction in considerably lower than that of the non-substituted one [90]; under standard polycondensation condition in acid medium, this leads to structures of the type

$$-CH_2-NH-C-NH- \\ \quad\quad\quad\quad\| \\ \quad\quad\quad\quad O$$

Scheme 12b

However, no matter how rigidly the matrix controls the structure of the "daughter" chain, the latter will always contain a fragment or fragments formed independently of the matrix, i.e., before its "recognition".

For matrix polymerization, the fraction of such fragments depends on the correlation between the length of the matrix and of the "daughter" chain. If the length of the "daughter" chain corresponds to that of the matrix, as, for example, for the polymerization of acrylic acid on polyvinylpyrrolidone [91], then the "daughter" chain will consist of two blocks (Scheme 13a).

Having reached the end of the matrix, the "daughter" chain goes on growing "in freedom" and the newly formed free fragments, having grown up to a certain length, are capable of "recognizing" a new chain of the matrix, and so on. This may be repeated many times, as, for instance, for AA and MA polymerization on PEO [76]; in such cases, the "daughter" chain will consist of alternating blocks which have grown without and under control of the matrix (Scheme 13b).

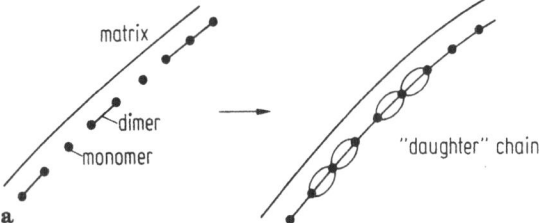

The less stable the matrix-"daughter" chain polycomplex is, i.e. the lesser the value K_1 in Eq. (12) is, the less will be, all other conditions being equal, the fraction of fragments which grow under matrix control. Changes in the conditions of the polymerization process lead to changes in the fraction of these fragments if the stability of the polycomplex (i.e. the capability of the growing chain for "recognizing" the matrix) depends on the conditions [1].

Under matrix polycondensation considerable influence exerted by the stability of the polycomplex of matrix and growing chains also is possible if the direction (mechanism) of the elementary step of the polycondensation depends on where this step takes place — in solution or at the matrix (i.e., with participation of polymer, oligomer and even monomer molecules adsorbed on the matrix). Scheme 14 shows cases corresponding to strong 14a and weak 14b binding of growing chains in a polycomplex:

Scheme 14a

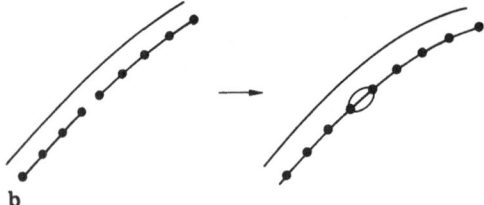

b   Scheme 14b

The circles here denote bonds which have been formed under matrix control.

If the matrix is "recognized" by very short growing chains, so that even monomer molecules and very short oligomers can be held back by the matrix (*left-hand* side of the Scheme 14a), then the greater part of the elementary growth steps are matrix-controlled (*right-hand* side of Scheme 14a). If the stability of the matrix-"daughter" chains polycomplex is weak, with only sufficiently long growing chains being capable of "recognizing" the matrix (see *left-hand* side of Scheme 14b), then the latter controls only a small fraction of the elementary growth steps of the "daughter" chain, and its structure practically does not differ from that of chains which have grown in the absence of the matrix (Scheme 14b, *right-hand* side). The matrix effect in the latter case may be manifested in the appearance of a fraction of high molecular weight "daughter" polymer because adsorption of the longest growing chains on the matrix may lead to a sharply increased probability of their condensation with each other. MWD of the polymer formed in the presence of matrix in this case is expected to be bimodal.

It is by transition from Scheme 14a to Scheme 14b that the authors of works [88, 89] explain the influence of the pH value on the structure of the "daughter" chains for the matrix polycondensation of urea and formaldehyde in water solution in the presence of PAA. At pH < 3.7 the matrix polycondensation product is enriched in units of structure Scheme 12a as compared with PFU, obtained at the same pH but without the matrix, while at pH > 3.7, the contents of these units in both matrix and non-matrix PFU practically coincide corresponding to structure Scheme 12b. Being stabilized by H-bonds the polycomplex PAA-PFU is stable in the range of low pH values, i.e., in this range the matrix recognizes by shortest growing chains; in the range of pH (in this particular case at pH < 3.7) the realization of Scheme 14a is possible. With increased pH (in the system under discussion pH > 3.7) progressive decrease in the polycomplex stability due to ionization of PAA is equivalent to the increase in the length of the matrix-"recognizing" oligomer chains of PFU growing in the solution according to the standard mechanism, i.e., a decrease in the fraction of the elementary matrix-controlled growth steps Scheme 14b. Changes in the structure of the "daughter" chains along with changes in the reaction medium are reflected by the structure and physico-chemical properties of the matrix polycondensation products, the polycomplexes PAA-PFU, in particular by the potentiometric titration curves, the rate of intermolecular imidation in polycomplexes on heating, and the kinetics and the limiting value of swelling in water [92].

4.3 "Recognition" of the Strongest Matrix Under Polymerization in the Presence of more than One Matrix

If two or more matrices of different structure are simultaneously present in the reaction system, then the growth of chains is controlled only by that matrix whose polycomplex with these chains turns out to be more stable [12]. This happens because the chains of the polymer under formation, as they grow freely, are capable of "recognizing" this strongest matrix earlier than the other ones since "recognition" occurs when a lesser degree of polymerization has been reached. Even if these freely growing chains happen to reach the length sufficient for binding with the weaker matrix and "recognize" it, owing to the substitution reaction, they will, in the long run, find themselves bound with the strongest matrix. Thus, due to dependence (12), there practically exists in the reaction system one polycomplex consisting of the polymer under formation and the strongest matrix up until the latter has been exhausted (i.e., up until its complete binding in the polycomplex with the accumulated polymer).

Such a situation was observed for the polymerization of MA in the simultanuous presence of two matrices: PVPD + PEO, PVPD + PVA or PEO + PVA (Fig. 10) [12]. Each of the PVPD, PEO and PVA matrices is capable of controlling the rate of the "daughter" polymer formation (see curves 1–4). However, in the presence of two matrices, only one of them, the strongest one with regard to the stability of its polycomplex with the "daughter" chain (see Table 1), controls the growth of these chains; these range according to their strength factor as PVPD > PEO > PVA which is in good agreement with the data on reactions of macromolecular substitution in the systems containing such macromolecules with the same degrees of polymerization (see Table 1).

As soon as the strong matrix is completely exhausted in the process of polymerization, i.e., after it has bound in a polycomplex with the polymer under formation, the

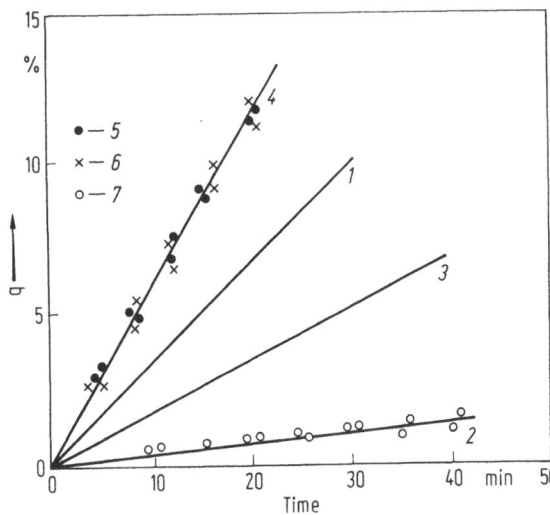

Fig. 10. Dependence of conversion on time for the polymerization of MA without matrix (*1*) and in the presence of matrices: PEO (*2*), PVA (*3*), PVPD (*4*), PVPD and PEO (*5*), PVPD and PVA (*6*), PEO and PVA (*7*) [12]

growing chains become capable of "recognizing" a weaker matrix which from that moment starts on controlling the polymerization process [12].

A similar phenomenon of "recognizing" the strongest matrix in the process of polymerization was observed in the system where the surface of micelles SAS (in the system MA-PVPD-polyethyleneglycolmonolaurate) was used as one of the matrices [93].

4.4 Matrix Regeneration in Processes of Matrix Synthesis of Polymers

Since the polycomplex is a product of matrix polymerization or polycondensation, each macromolecule of the matrix works as a matrix only once. All attempts to create systems with the workable "zip-up" mechanism of polymerization of monomer molecules, preliminarily bound with the matrix, with automatic liberation of the formed "daughter" chain from the matrix, have ended in failure (this was assumed to be possible, when, for instance, stabilization of the monomer-matrix complex takes place due to the presence of the double bond in the monomer [94,95].

Yet, in principle, it is possible to create systems where the liberation of the matrix from the polycomplex takes place in the process of the growth of the "daughter" chains, and hence the matrix is capable of working many times. This was theoretically demonstrated in the work of the authors of this review [96]. (We do not consider a trivial case when, the matrix having been exhausted, a stronger complex agent is introduced, and freeing of the matrix occurs due to its substitution in the polycomplex with this stronger complex agent).

Such reaction systems must include, besides monomers (or comonomers), relatively short chains of the strong matrix and long chains of the weaker matrix. Further, after the end of the growing chain has reached the end of the chain of the strong matrix, the chain-growth does not end, i.e., the growth of the "daughter" chain continues in accordance with Scheme 13 b.

These conditions having been observed, the following situation develops:

1. The growing chain "recognizes" the strong matrix; as long as the length of the "daughter" chain is less than that of the strong matrix chain n_1 (i.e., the active center has not moved away from the matrix), the stability of the polycomplex is characterized by K_{11}^n, and the condition $K_{11}^n \gg K_{12}^n$ is fulfilled where K_{11} and K_{12} correspond to K_1 in Eq. (12) for complexes of the "daughter" chain with a strong and the weak matrix, respectively (Scheme 15a).

2. The length of the "daughter" chain, as it grows, exceeds that of the strong matrix. Then with $n > n_1$, the stability of the polycomplex of this matrix with growing chain is characterized by $K_{11}^{n_1}$ and no longer depends on n. Since the length of the weaker matrix is $n_2 \gg n_1$, the stability of its polycomplex with the growing chain continues to increase in accordance with K_{12}^n (Scheme 15b). With length of the growing chain $n = n^*$, $K_{11}^{n_1} = K_{12}^n$ is achieved (Scheme 15c), and with $n > n^*$, already $K_{11}^{n_1} < K_{12}^n$ (Scheme 15d). This means that with $n \simeq n^*$, the stability of the polycomplexes of the growing chain with both matrices becomes equal, and with $n > n^*$, the weak matrix begins to substitute the strong one, liberating it from the polycomplex according to the inversion of the selection factor (see Eq. (22)). Thus, the regenerated macromolecules of the strong matrix will be ready

to catch new growing chains, and the "daughter" chain will continue to grow on the weak matrix.

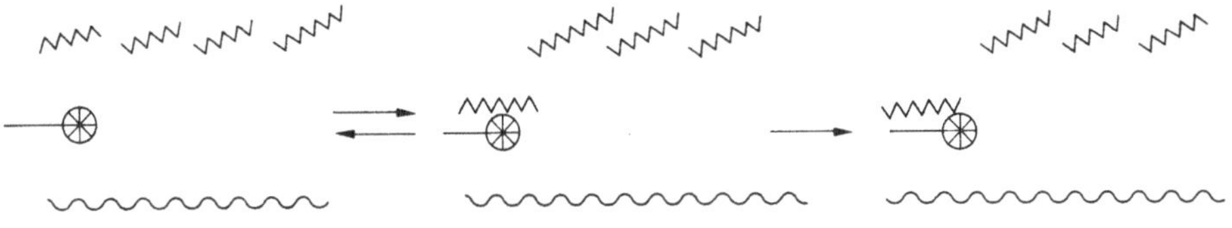

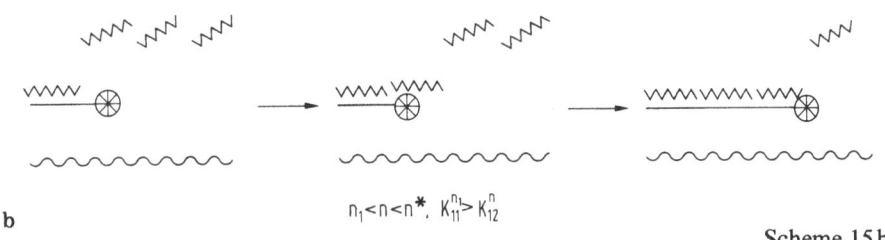

a
$$n \leq n_1, K_{11}^n \gg K_{12}^n$$
Scheme 15a

b
$$n_1 < n < n^*, K_{11}^{n_1} > K_{12}^n$$
Scheme 15b

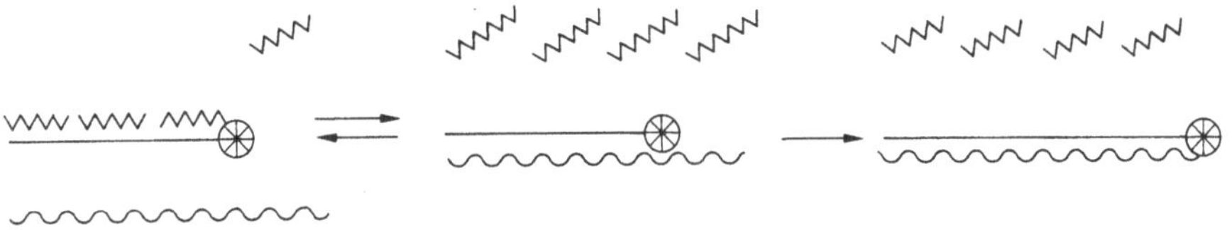

c
$$n \simeq n^*, K_{11}^{n_1} \simeq K_{12}^{n*}$$
d
$$n > n^*, K_{11}^{n_1} < K_{12}^n$$
Scheme 15c
Scheme 15d

By varying the chain lengths of the strong matrix, all other conditions being equal, chain growth of the "daughter" polymer is controlled either by practically only the strong matrix, or the weak one, or by both; in the latter case, the resulting "daughter" chain contains commensurable blocks grown on both matrices. The rate of polymerization will correspond to the rate of polymerization on the strong or on the weak matrix, or will have an intermediary value [96].

A similar picture was experimentally observed for the polymerization of MA in the presence of PVPD macromolecules and micelles of polyethylenglycolmonolaurate: at high degrees of polymerization of the strongest matrix PVPD, the polymerization process was controlled only by this matrix, with low ones by both [93].

5 Conclusion

We have confined this review only to interpolymer interactions although it is obvious that a number of theoretical conclusions concerning such interactions, are, at least qualitatively, applicable to interactions of macromolecules with the surface of different types of particles.

It is much simpler to calculate an interpolymer interaction which is sufficiently well modelled by adsorption of chains on a one-dimensional lattice, than to calculate adsorption on the two-dimentional one as well as to calculate adsorption of small particles on the macromolecule. When calculating adsorption on the two-dimensional lattice (usual surface), for example, difficulties rapidly pile up with increasing the degree of filling of the surface with macromolecules; it is difficult to take into consideration self- and intercrossings of the adsorbed macromolecules, the influence exerted by the length and rigidity of these macromolecules, etc.

Nevertheless, there are definite grounds to believe that under adsorption of macromolecules on the "infinite" two-dimentional lattice (surface), the dependence of the adsorption constant on the chain length, with low filling degrees of the surface and not too large chain length, will be expressed by an equation close to (12) with all the following consequences in regard to selectivity of adsorption and substitution of adsorbed macromolecules with other ones.

The above developed assumptions may also be applied to the interactions of macromolecules with small particles, in particular with globulae, particles of sols, if the length of the macromolecule is sufficiently high to be capable of binding simultaneously with more than one particle.

Thus, if the macromolecule forms only one loop around the particle, then the quantity of the bonds of the particle with the macromolecule is proportional to the length of the particle's circumference, and the correspondent constant of equilibrium of the association-dissociation depends on the radius of particles in accordance with

$$K_r = K_1^r \tag{25}$$

but if the macromolecule "coils itself" around the particle closing practically its whole surface, then the quantity of the bonds of the particle with the macromolecule is proportional to the surface of the particle, and

$$K_r = K_1^{r^2} \tag{26}$$

where K_1 has the same meaning as in Eqs. (8, 12). In the rest of the cases, the exponent in the equation for K_r will have an intermediary meaning somewhere between r and r^2.

Equations (25) and (26) give grounds to expect that, for instance, dispersions of very small particles can be fractionated by size with the help of macromolecules which are capable of interacting with them.

It goes without saying that many assumptions, developed for matrix polymerization and polycondensation and connected with "recognizing" the matrix and the growing chain, may be applied to the corresponding reactions which run on the surface.

The authors of this review have to confess that it was a very difficult task to suppress the temptation of including extensive literary source dealing with "recognition" in systems with biologic macromolecules or their synthetic analogues. The sole reason for the authors' apprehension was that the reader, unacquainted with the field of interpolymer interactions, might be under the impression that the systems of simple synthetic macromolecules in some aspects, serve as no more than simplified models of the systems of biological macromolecules. In reality it was the intention of the authors to show that molecular "recognition" is a consequence of the polymer nature of the reagents and a fundamental property of macromolecular systems, as well as to demonstrate some effects of this fundamental property.

6 References

1. Kabanov VA, Papisov IM (1979) Vysokomolek soed, A21: 243
2. Tsuchida E, Abe K (1982) Adv. Polym. Sci. 45: 1
3. Philipp B, Dawydoff W, Linow K (1982) Z. Chem., 22 Jg: 1
4. Bamford CH (1982) Chem. Austral 49: 341
5. Kabanov VA, Zezin AB (1984) Makromol. Chem. Suppl. 6: 259
6. Zezin AB, Rogacheva VB (1973) In: Uspehi khimii i fiziki polimerov, Khimiya, Moscow, p 3
7a. Bekturov EA, Bimendina LA (1979) Interpolimernye kompleksy, Nauka, Alma-Ata
7b. Bekturov EA, Bimendina LA (1981) Adv. Polym. Sci., 41: 99
8. Gladilin KL, Orlovsky AF, Kirpotin DB, Oparin AI Origin of life 1978: 357
9. Oparin AI, Gladilin KL (1980) Bio Systems 12: 133
10. Oparin AI, Gladilin KL (1980) Uspehi biol. khimii, 21: 3
11. Papisov IM, Nedyalkova TsI, Avramchuk NA, Kabanov VA (1973) Vysokomolek. soed. A15: 2003
12. Papisov IM, Nekrasova NA, Pautov VD, Kabanov VA (1974) Doklady AN SSSR 214: 861
13. Abe K, Koide M, Tsichida E (1977) Macromolecules 10: 1259
14. Baranovsky VYu (1987) In: III Vsesoyuznay konferentsiya Vodorastvorimy polimery i ih primenenie, Irkutsk, p 126
15. Korugic-Perković L, Ferguson J (1983) Polymeri (SFRI) 4: 301
16. Chatterjee SK, Sethi KR (1984) Polymer 25: 1367
17. Izumrudov VA, Bronich TK, Zezin AB, Kabanov VA (1984) Doklady AN SSSR 278: 404
18. Litmanovich AA (1978) Vestnik Moskovskogo Universiteta, seriya khimiya 19: 617
19. Litmanovich AA, Papisov IM, Kabanov VA (1980) Vysokomolek. soed., A22: 1180
20. Litmanovich AA, Papisov IM, Kabanov VA (1981) Eur. Polym. J. 17: 981
21. Litmanovich AA, Anufrieva EV, Papisov IM, Kabanov VA (1979) Doklady AN SSSR, 246: 923
22. Kikuchi Y, Kubota N (1985) Makromol. Chem. Rapid Commun. 6: 387
23. Izumrudov VA, Bronich TK, Saburova OS, Zezin AB, Kabanov VA (1986) Vysokomolek. soed. B28: 725
24. Applequist J, Dumle V (1965) J. of ACS, 87: 1450
25. Dumle V (1970) Biopol., 9: 353
26. Papisov IM, Baranovsky VYu, Sergieva EI, Antipina AD, Kabanov VA (1974) Vysokomolek. soed., A16: 1133
27. Papisov IM, Litmanovich AA (1977) ibid, A19: 716
28. Baranovsky VYu, Litmanovich AA, Papisov IM, Kabanov VA (1981) Eur. Polym. J. 17: 969
29. Khodakov YuS, Berlin AA, Kalyaev GI, Minachev KhM (1969) Teoreticheskaya i eksperimental'naya khimiya 5: 631
30. Magee W, Gibbs J, Zimm B (1963) Biopol. 1: 133

31. Magee W, Gibbs J, Newell R (1965) J. Chem. Phys. 43: 2115
32. Tsuchida E, Osada I (1971) Makromol. Chem. 175: 593
33. Bresler SE, Chernajenko VM, Saminski EM (1972) Biopol., 11: 1541
34. Hill T (1960) Introduction in statistical thermodynamics, Addison-Wesley Reading Mass.
35. El'yashevich AM, Skvortsov AM (1971) Molekulyarnaya Biologiya 5: 204
36. Papisov IM, Baranovsky VYu, Chernyak VYa, Antipina AD, Kabanov VA (1971) Doklady AN SSR 199: 1364
37. Baranovsky VYu, Papisov IM (1974) ibid. 217: 123
38. Birshtein TM, El'yashevich AM, Morgenshtern LA (1972) Vysokomolek. soed., B14m487
39. Birstein TM, Eliashevich AM, Morgenstern LA (1974) Biophys. Chem., 1: 242
40. Lutsenko VV, Lopatkin AA, Zezin AB (1974) Vysokomolek. soed A16: 2429
41. Lutsenko VV, Zezin AB, Kalyugnaya RI (1974) ibid A16: 2411
42. Litmanovich AA, Kazarin LA, Papisov IM (1976) ibid B18: 681
43. Kazarin LA, Baranovsky VYu, Litmanovich AA, Papisov IM (1983) ibid, B25: 212
44. Papisov IM, Bolyachevskaya KI, Litmanovich AA, Markov SV, Baranovsky VYu, Kazarin LA (1983) IUPAC MACRO '83, Bucharest, Abstr. Sec. 6–7, s. a. p 96
45. Baranovsky VYu, Kazarin LA, Litmanovich AA, Papisov IM (1984) Eur. Polym. J. 20: 191
46. Bolyachevskaya KI, Litmanovich AA, Papisov IM, Litmanovich AD, Cherkezyan VO (1985) Vysokomolek. soed., B27: 494
47. Saitbaeva SS, Bimendina LA, Bekturov EA (1979) Izvestiya AN KazSSR, serya Khimiya
48. Antipina AD, Baranovsky VYu, Papisov IM, Kabanov VA (1972) Vysokomolek. soed., A14: 941
49. Bailey F, Lundberg R, Gallard R (1964) J. Polym. Sci., A2: 845
50. Smith K, Winslow A, Petersen D (1959) Industr. Eng. Chem., 51: 1361
51. Blake R (1972) Biopol. 11: 913
52. Litmanovich AA, Kirsh YuE, Papisov IM (1978) Vysokomolek. soed., B20: 83
53. Kokufuta E, Yokota A, Nakamura I (1983) Polymer 24: 1031
54. Kharenko OA, Kharenko AV, Kasaikin VA, Zezin AB, Kabanov VA (1979) Vysokomolek. soed. A21: 2726
55. Kabanov VA, Zezin AB, Rogacheva VB, Righikov SV (1982) Doklady AN SSSR, 267: 862
56. Fraktsionirovanie polimerov (1971), Moscow, Mir, p 321
57. Rosental AJ, White BB (1952) Industr. Engng. Chem. 44: 2693
58. Litmanovich AA (1985) Vysokomolek. soed. B27: 350
59. Aleksina OA, Zezin AB, Papisov IM (1971) ibid A13: 1199
60. Kabanov VA, Evdakov VP, Mustafaev MI, Antipina AD (1977) Molekulyarnaya biologiya, 11: 582
61. Ermakova LN, Frolov YuG, Kasaikin VA, Zezin AB, Kabanov VA (1981) Vysokomolek. soed. A23: 2328
62. Anufrieva EV, Pautov VD, Papisov IM, Kabanov VA (1977) Doklady AN SSSR 232: 1096
63. Chen H-L, Morawetz H (1982) Macromolecules 15: 1445
64. Chen H-L, Morawetz H (1983) Eur. Polym. J. 19: 923
65. Izumrudov VA, Savitskii AP, Zezin AB, Kabanov VA (1983) Vysokomolek. soed. B25: 805
66. Izumrudov VA, Savitskii AP, Zezin AB, Kabanov VA (1983) Doklady AN SSSR 272: 1408
67. Izumrudov VA, Savitskii AP, Zezin AB, Kabanov VA (1984) Vysokomolek. soed. A26: 1724
68. Izumrudov VA, Bakeev KN, Kabanov VA (1986) In: 3 Vsesoyuznaya konferentsya po khimii oligomerov, Odessa: Tezisy plenarnyh i stendovyh dokladov, Chernogolovka, 1986, p 171
69. Izumrudov VA, Bakeev KN, Zezin AB, Kabanov VA Doklady AN SSSR 1986: 1442
70. Izumrudov VA, Zezin AB, Kabanov VA ibid 275: 1120 (1984)
71. Izumrudov VA, Savitskii AP, Bakeev KN, Zezin AB, Kabanov VA (1984) Makromol. Chem. Rapid Commun. 5: 709
72. Izumrudov VA, Bronich TK, Zezin AB, Kabanov VA (1987) Vysokomolek. soed., A29: 1224

73. Kuchanov SI, Bakeev KN, Izumrudov VA, Zezin AB, Kabanov VA (1987) In: III Vsesoyuznaya konferentsia Vodirastvorimye polimery i ih primenenie, Irkutsk p 71
74. Oparin AI, Gladilin KL, Kirpotin DB, Chertihin GB, Orlosky AF (1977) Doklady AN SSR 232: 485
75. Osada E, Antipina AD, Papisov IM, Kabanov VA, Kargin VA (1970) ibid, 191: 399
76. Papisov IM, Kabanov VA, Osada E, Leskano-Brito M, Reimont J, Gvozdetsky AN (1972) Vysokomolek. soed. A14: 2462
77. Nedyalkova TsI, Baranovsky VYu, Papisov IM, Kabanov VA (1975) ibid B17: 174
78. Anufrieva EV, Gotlib YuYa, Krakovyak MG, Skorohodov SS (1972) ibid A14: 1430
79. Papisov IM, Sergieva EI, Pautov VD, Kabanov VA (1974) Doklady AN SSSR, 208: 123
80. Illiopoulos I, Audebert R (1985) Polymer Bull., 13: 171
81. Kabanov VA, Kargina OV, Petrovskaya VA (1971) Vysokomolek. soed. A13: 348
82. Smid J, Tan YY, Challa G (1983) Eur. Polym. J. 19: 853
83. Smid J, Tan YY, Challa G (1984) ibid, 20: 887
84. Osada E, Nekrasova NA, Papisov IM, Kabanov VA (1970) Vysokomolek. soed., B12: 324
85. Polowinski S (1983) Eur. Polym. J. 19: 679
86. Polowinski S (1984) J. Polym. Sci., Polym Chem. Ed., 22: 2887
87. Kargina OV, Mishutina LA, Soshinsky AA, Kabanov VA (1984) Doklady AN SSSR 275: 657
88. Papisov IM, Kuzovleva OE, Markov SV, Litmanovich AA (1984) Eur. Polym. J. 20: 195
89. Litmanovich AA, Papisov IM, Markov SV (1984) Doklady AN SSSR 278: 676
90. Entsiklopediya polimerov, Moscow, Sovetskaya entsiklopediya, (1974) vol 2 p 311
91. Ferguson J, Al-Alawi S, Granmayeh R (1983) Eur. Polym. J. 19: 475
92. Litmanovich AA, Markov SV, Papisov IM (1986) Vysokomolek. soed. A28: 1271
93. Baranovsky VYu, Gnatko NN, Litmanovich AA, Papisov IM (1988) ibid (in press)
94. Tsuchida E, Osada Y (1975) J. Polym. Sci., Polym. Chem. Ed. 13: 559
95. Papisov IM, Garina ES, Kabanov VA, Kargin VA (1969) Vysokomolek. soed. B11: 614
96. Papisov IM, Litmanovich AA (1985) ibid A27: 2157

Editor: H. Höcker
Received December 27, 1988

Author Index Volumes 1—90

Subject Index